Energy Efficiency and Conservation in Metal Industries

Energy Efficiency and Conservation in Metal Industries

With Selected Cases of Investment Grade Audit

Lt. Swapan Kumar Dutta
Jitendra Saxena
Binoy Krishna Choudhury

CRC Press
Taylor & Francis Group
Boca Raton London

CRC Press is an imprint of the
Taylor & Francis Group, an **informa** business

First edition published 2023
by CRC Press
6000 Broken Sound Parkway NW, Suite 300, Boca Raton, FL 33487–2742

and by CRC Press
4 Park Square, Milton Park, Abingdon, Oxon, OX14 4RN

ISBN: 978-0-367-74324-6 (hbk)
ISBN: 978-0-367-74327-7 (pbk)
ISBN: 978-1-003-15713-7 (ebk)

DOI: 10.1201/9781003157137

Typeset in Times LT Std
by Apex CoVantage, LLC

We seek blessings from our parents and the departed soul of our senior-most member, the late Prof Swapan Kumar Dutta, to whom we dedicate this book.

Contents

Preface

As rapid transformation takes place in the energy marketplace and energy cost emerges to be the most significant manageable cost among man/material/machine costs of production, the role of energy management and conservation has become prominent for developed and developing countries in both their efforts toward decarbonization while making their industry sectors globally more competitive. Reducing utility costs is one of the most effective and achievable strategies for lowering operating costs, as well as making industries more competitive while helping save the local and global environment. While energy efficiency improvement technologies have been available for decades, scaling up their implementation faces numerous challenges, particularly in emerging economies like India. There is a need for shifting the needle of the energy management and conservation debate from the shop floors to the board rooms. In India, the Perform, Achieve and Trade (PAT) Scheme under the National Mission for Enhanced Energy Efficiency (NMEEE) has established an enabling policy environment through the Energy Conservation Act, 2001 implemented under the leadership of the Bureau of Energy Efficiency of the Ministry of Power, that effectively promotes the energy efficiency improvements in the largest industry sectors in India through an innovative market-based incentive mechanism.

Energy efficiency improvements in energy-intensive sectors like metal industries are going to be an increasingly important area of focus for the next several decades in our effort to achieve a net zero greener future. Today, many large energy consumers are contracting with energy efficiency equipment suppliers and energy service companies (ESCOs) and other service providers to implement clean energy projects. The ESCO arrangement allows a market-driven approach to financing and implementing energy management in industries, buildings, municipalities, etc., through performance-based energy savings contracting modalities. The success of these mechanisms to scale up the impact is anchored around the establishment of robust baseline energy consumption and the quantification, measurement and verification (M&V) of energy savings resulting from project implementation. Upstream accurate and comprehensive energy audits are essential as a means to assess and verify a project's success at meeting contracted goals. This book, by leading practitioners and experts, is designed to provide you with the fundamental knowledge you need to evaluate how energy is used in metal-industrial facilities, establish accurate baseline information and identify where energy consumption can be reduced. The book will also provide you with all the information you need to establish an energy audit program for your facility. Basics of energy management and conservation in metal industries provide overall ideas of the industry, including all aspects of energy auditing—including metallurgical, mechanical and electrical aspects. This book is a first-level metal industries energy audit reference for energy engineers, as well as non-engineers and others new to the field of energy management. It will guide the reader through the energy audit process system to help them identify and prioritize conservation potential, as well as to identify several low-cost and no-cost energy efficiency improvement and conservation and retrofitting opportunities in metal industries. The book provides case studies to help readers to understand and apply basic energy-saving and conservation measures to their energy projects. Real-life situations and practical experience have been given priority for productive use of the book. For example, 29 case studies on metal industries' energy efficiency/policy from countries around the globe, numerous figures and tables will be interesting to the readers of this book. By using this book as reference, energy consultants, auditors and managers can develop a methodology of optimum and effective use of energy with special focus on metal industries to make valuable contributions to ongoing and future efforts to improve energy efficiency.

Dr Ashok Sarkar, Senior Energy Specialist
and Task Team Leader in the Global
Energy Practice of the World Bank,
Washington DC, USA

Acknowledgments

Our special thanks to CRC Press, in particular Dr. Gagandeep Singh and his team and APEX, in particular Mr. Ganesh Pawan Kumar Agoor and his team, for the professional excellence and team spirit that have enlightened and invigorated us throughout this book project activities, and Dr. Gopal C. Dutta Roy, our relatives, friends and supporters from the engineering fraternity for their continuous encouragement and motivation and their hope that this reference book will be a source of information for the engineering community across the globe in the field of energy conservation and management with special focus on metal industries. We remain indebted to Dr. Ashok Sarkar, senior energy specialist and task team leader in the Global Energy Practice of the World Bank, Washington DC, USA and manager of World Bank's US$43 million Partial Risk Sharing Facility for Energy Efficiency (http://prsf.sidbi. in) project with SIDBI, and US$300 million Energy Efficiency Scale-up Program operation with EESL in India, for going through the abstracts of all the eight chapters and writing a preface/foreword to this book.

Our heartfelt thanks to President Mr. Vonderleyan and to Ms. Mariana Kotzeva Director General of Eurostat, European Commission and Mr. Maciej Grzeszczyk, Policy officer of European Commission, for giving us the permission to republish data and figures on EE&C in metal industries. Our sincere thanks and gratitude for the permission from Mr. Ajay Bakre, Director General, Bureau of Energy Efficiency (BEE), Ministry of Power, Government of India and Mr. Arijit Sen Gupta, Director (BEE) New Delhi for using all study materials of BEE including the report on "Improving Energy Efficiency in Iron and Steel Sector under Perform Achieve and Trade Scheme of Bureau of Energy Efficiency," September 2018, published by BEE in partnership with GIZ India German Energy Programme.

Last but not least, we also acknowledge Dr. Gaurav Pandey, senior post-doctorate researcher from India, for his unparalleled academic support, and Sri Rhitambhar Choudhury, student of Class XII at South Point High School, Kolkata, for the cover.

Authors

Swapan Kumar Dutta (Deceased) successfully founded and managed various business ventures in engineering design, manufacturing and consulting in a career spanning more than 30 years. He started his career as a design engineer at International Combustion India. Thereafter, he founded Yantra Shilpa Udyog and Hitech Equipments, leading Indian designers, manufacturers and exporters of complex engineering systems used for conveying, bulk materials handling, mineral processing and specialized ceramics. He was also the founder and principal consultant of an energy consulting firm providing advice on energy conservation and energy efficient design.

Mr. Dutta earned undergraduate (1981) and Master's (1983) degrees in mechanical engineering, both from Jadavpur University. He completed executive training in quality management (IRCA, UK), quality management (AOTS, Japan), energy management and audit (Jadavpur University), and advance environment management system (EARA, UK). He was an internationally accredited energy auditor and an accredited management teacher of management of public utility System—AIMA-CMS, and a faculty member of the Energy Management Department of the Indian Institute of Social Welfare and Business Management (IISWBM), Kolkata.

Jitendra Saxena Ph.D, a certified energy manager from Bureau of Energy Efficiency Ministry of power Government of India and an engineering professional having vast experience in the field of academic and industry, joined The Institution of Engineers (India) on 9 September 2016 as director (education, examination and accreditation). He has more than 25 years of experience in teaching, technical education, planning and management, project management and industries. After obtaining AMIE degree from IEI, he did his M.Tech in the field of energy management from Devi Ahilya Vishwavidyalaya, Indore. Thereafter, he obtained his doctorate from Maulana Azad National Institute of Technology, Bhopal. He started his modest career in industries and switched over to technical education. Gradually, he rose to the level of director of a technical institution imparting courses of both B.Tech and M.Tech. He also served at the Military College of Telecommunication Engineering (MCTE) as visiting professor. He was associated with various technical and industrial activities of PHD Chamber of Commerce and Confederation of Indian Industry (CII) in various capacities. He was also involved with energy and electrical safety auditing and consultancy in multinational companies and for designated consumers. As a result, experience is a perfect blend of academic as well as industry backgrounds. He is a chartered engineer from The Institution of Engineers (India), electrical supervisory safety certificate from Chief Engineer Electrical Safety Department, Government of Madhya Pradesh, India. He is a fellow member of IEI and IETE and a life member of Solar Energy Society of India and the Indian Society for Technical Education (ISTE).

Binoy Krishna Choudhury Ph.D., is a mechanical engineering graduate from Jalpaiguri Government Engineering College, received his Ph.D. from the Indian Institute of Technology (IIT) Kharagpur. He is a fellow of The Institution of Engineers (India), Association of Engineers (India) and Indian Water Works Association. He is also currently a professor in the energy management stream of the public systems management (PSM) program. He has more than 27 years of teaching/professional experience, which includes more than 24 years at the present Institute. He served as principal investigator and coordinator for more than 15 renewable energy and energy efficiency projects, which covered community development to industrial profitability and sustainability issues. He is an honorary vice president, for the Asia subcontinent in the Association of Energy Engineers, USA; member of the Energy and Environment Standing Committee, the Bengal National Chamber of Commerce and Industry and The Bengal Chamber; JICA Alumni Association of India; and Low-Carbon Energy for Development Network (UK). He is the founder and editor of *Energy Window*, is on the editorial board of *Reason: A Technical Journal* and reviewer of ACTA Press, Energy for Sustainable Development.

He secured a National Scholarship in 1981, graduated in 1988 (JGEC, North Bengal University) and earned his Ph.D. in 1994 (IIT, Kharagpur); he won a best research paper award in 1998 (AIMS, New Delhi) and is a winner of the Coal India (J.G. Kumara Mangalam Memorial) Award 2018. He is a certified energy auditor (CEA), certified energy manager (CEM) and trainer (CEA & CEM) of the Association of Energy Engineers (AEE), USA; MNRE empaneled national-level evaluator of solar thermal programs and accredited energy auditor (AEA) of the government of India. He served on the United Nations Economic and Social Commission for Asia and the Pacific (ESCAP) as economic affairs officer in the Energy Resources Section and contributed to the regional energy scenario assessment, improvement of energy efficiency and CDM workshop. He has contributed/published, more than 55 books, journal articles, international and national seminars, workshops, etc. He is the founding secretary of AEE India Chapter and Energy Club, and is life member of the Solar Energy Society of India, AEE, Indian Society for Technical Education, Ramakrishna Mission Institute of Culture and Fellow of Institution of Engineers (India), and Indian Water Works Association (IWWA).

1 Energy Efficiency and Conservation

Jitendra Saxena

1.1 HISTORICAL PERSPECTIVE

Ferrous alloys' most vital element is steel, which was produced in large quantities and at cheap costs during the 19th century. There are seven stages in which pig iron smelting and steel production was followed by advanced technology in iron and steel during World War I in 1914, then the negativity of the post-World War I years of the 1920s and the declining curve of pig iron output in major producing countries declining by 75% as compared to the pre-crisis peaks. Then driven demand (1940–1945) of war revived the output as the United States, the USSR, Germany and Japan reached new dimension in steel production records [1]. Then after World War II, direct reduction of iron (DRI) processes represent a basic iron-making innovation process and operating at present in 24 countries. DRI iron and steel techniques have not fulfilled their target by 2015; only 5% of the global iron output are produced by direct reduction iron plants. There are two technologies in steel-making processes: one is basic oxygen furnaces (BOFs), and the second is continuous casting, which have revolutionized the industry through waste reduction, increasing productivity and higher efficiency. Then during the 1970s, oil prices rose exponentially, resulting in increase of the cost of energy, which effect social, political and economic global condition at that point of time. All foremost consumers of electricity and fuel began to reduce their energy consumption and to lower the final energy intensity of their products [2]. As a result, the iron and steel industry has been a significant contributor of these tedious works. The history of iron making can be seen for high energy efficiency, and the continuous rigorous effort brought typical fuel requirements with 200 GJ/t of pig iron in 1800 to less than 100 GJ/t by 1850, to only about 50 GJ/t by 1900 (Heal, 1975), and to less than 20 GJ/t after a centenary.

1.2 METAL INDUSTRIES: LOCAL AND GLOBAL PERSPECTIVES

There are training programs and courses on energy management, but training and skill sets for applications in the materials and metal industry are limited. The energy used for lighting in metals plants normally lies between 2% and 5% of the total energy consumption. The industrial sector is the largest consumer of electrical energy, and the transport sector is the second largest. The global information discusses energy usage for HVAC (heating, ventilation, and air conditioning); many metals plants have specific heating and air conditioning. In metal industries, process

DOI: 10.1201/9781003157137-1

heating is the most vital area of energy consumption. The potential of energy efficiency requirement programs remains unchanged. Since 2014 at distinct level of targeting, efficiency in investment remains unaltered. The materials industry must use a huge amount of energy to melt and for heat treatment of metal. Energy costs are normally a major portion of costs (labor and metal costs are also high). The primary metal industry uses 8% of all energy in manufacturing industries for processing. It has been estimated that energy use in the iron and steel industry can be reduced by 39% with commercially available technologies, and another 24% with new and innovative technologies [3]. For aluminum it is estimated to be 33% savings with current technologies and another 30% with new technologies [4].

Energy-saving techniques which are universally adopted across the globe for metal industry include the following:

1. Coke drying and its quenching mechanism
2. Recovery of waste heat from sintering, rolling and casting processes
3. Coal moisture control
4. Coke-oven program heating
5. Injection and pulverization of coal in blast furnaces
6. Top gas recovery turbine
7. Stove gas waste heat recovery
8. Enrichment of oxygen in stove gas
9. Sensible heat recovery from stove gas
10. Ladle heating with programmable logic control
11. Recovery of basic oxygen furnace gas
12. Recuperative burners for reheating
13. Energy analytics with energy informatics and control systems

1.3 ENERGY CONSERVATION GLOBAL IMPERATIVES

Global imperatives with special reference to energy conservation depend upon primary energy consumption and energy intensity as the most dominating factors with gross domestic product (GDP) growth rate. Primary energy intensity is a vital point which reveals how much energy is utilized by the global economy. Climatic conditions play a vital role for also developed countries also, according to climate conditions for heating and cooling (EEA, 2014; EC, 2014; IEA, 2014a, 2014b, 2014c; World Energy Outlook, 2013, 2014; World Bank, 2014,COM 2014). In winter and cold season, the gas demand for heating is increased by 2% and there is improvement in energy intensity up to 1.4% in the year 2017. When discussed with respect to input supply chain, coal power generation was increased by 3% and 2.5% in 2017 and 2018, respectively, which gives electric supply for sturdy electrical demand growth. There are different categories of targeting efficiency of investment which remain unchanged from 2014. Industrial and transport sectors were about 0.6% larger in 2018 in comparison to 2017. The instability in the oil market will be overcome by technical efficiency and its requirement to reduce oil imports in the world global economies tentatively equal to 165 million tons of oil equivalent (MTOE), which is equal to total annual oil demand (primary) of Belgium, Australia and Germany (Eurostat &OECD, 2005; John Williamson, 1983; Chugoku, 2000; Noriki Hirose, 2005; IEA, 2017). There is a significant reduction of 20% in oil imports due to continuous energy improvements since 2000. Globally, the world will be more environmentally friendly and will be closer for sustainable emission pathway and will be highly consistent for climate objectives and goals. Energy conservation is an important factor and it is internationally recognized in the world's developed and developing countries. Japan is the most predominant country for energy conservation measures and energy reduction technology. Depletion of coal is likely to be there in the coming years, which take three million years to form. The consumption of fossil fuels is of 60%, and therefore, there is a high requirement for energy efficiency conservation and its measures. Energy conservation is directly related to energy consumption. If we increase energy consumption, energy conservation will reduce, so it is very important to reduce energy consumption for energy conservation measures and techniques. Energy conservation is the result of series of processes, and its developments such as technology developments or productivity increase. Energy efficiency is a contributor and integral part of energy conservation. Energy efficiency and conservation (EE&C) technologies are explained in detail in Chapter 3 highlighting waste heat recovery (WHR) and combined heat and power (CHP) systems. Energy intensity and GDP should be matched with the global economy. Energy efficiency is obtained when areas of production of specified product or consumption is reduced without sacrificing the comfort level (Lawrence Berkeley National Laboratory [LBNL], 2008).

A key promotional area of energy efficiency will provide significant measures to energy conservation and it validate the integral part of promotional policies of energy conservation and its initiatives. Energy efficiency is very closely related with environmental factors like climate change and its measures. The most cost-effective and reliable means of mitigating the global climate change is the control of CO_2 emissions through energy efficiency improvements. Chapter 2 and Chapter 8 provide in-depth studies of environmental factors related to societal growth and sustainability. Energy efficiency means that a lesser amount of energy is utilized to complete the same process or function.

Important key points for energy conservation are the following:

1. Reduced energy consumption and its tariff costs (European Commission, EC 2014 b, c, d, e, f)
2. Increasing healthy competition
3. Increasing production
4. Increasing quality and total quality management (TQM)
5. Increasing profit margin through energy efficiency measures
6. Conservation of natural resources
7. Energy security
8. Reduction of greenhouse gases (GHG) and hazardous emissions
9. Sustainability and sustainable environment

The above key indicators to provide sustainable growth nationally and globally, and will provide a platform for sustainability. By adopting energy conservation measures and optimum utilization of resources, the GDP growth rate of the country will increase, and purchasing power will also increase.

1.4 ENERGY EFFICIENCY AND CONSERVATION: FUTURE TRENDS AND SCOPE

For fossil fuels and different grades of coal which took a large span of time to form—nearly three million years—their depletion time is too short. The consumption of natural resources is exponentially increasing, and we have consumed 60% of natural resources. The rate of consumption is more than creation or formation of natural resources. Natural balance of fossil fuels is disturbed due to high dependency on them. In order to maintain sustainability and for sustainable development, energy efficiency measures should be universally adopted across the globe.

Energy conservation and energy efficiency are two distinct energy-specific terms, but are correlated concepts for energy efficiency measures and judicious utilization of energy. Energy conservation is measures in physical sense and it is directly related to energy consumption reduction technique. Energy efficiency is energy intensity in a specific process, product or specific area of consumption or production is reduced without affecting consumption, comfort levels or output. Energy efficiency promotional activities and policies implementation will add to energy conservation and is therefore a central part of energy conservation, which is discussed in detail in Chapter 2 with different energy technologies, processes and energy audits in metal industry in Chapter 3, Chapter 4 and Chapter 5, respectively. If the process or production system is energy efficient, it will be more resourceful with reducing pollution and efficient ecosystem. (EC 2014 g, h, i. j, k, COM 2013).

If traditional light bulbs are replaced with compact fluorescent lamps (CFL) bulbs, energy usage would be one-fourth of the energy to light a room. Pollution levels reduce by the same amount, and it is proven that CO_2 emission for incandescent lamp is 63 g/hr whereas for fluorescent lamps, CO_2 emission is 16 g/hr [5]. LEDs are very efficient relative to every lighting type on the market and extremely efficient relative to incandescent bulbs. Typical source efficiency ranges 37 and 120 lumens/watt. Most values for LED system efficiency fall above 50 lumens/watt.

After the first oil crisis in 1973, the energy efficiency concept drew global attention, and at the present time, it is the most vital and cost-effective solution for mitigation of climate change and to reduce carbon footprint. The industrial sector accounts for some 41% of global primary energy demand and approximately the same share of CO_2 emissions.

The most important features of energy efficiency include the following:

- Energy bills reduction and minimization
- Healthy and technical competition
- Increased in production
- Quality of service
- Profit maximization
- Reduction in imports of energy so duty will be reduced

- Conservation of limited resources
- Energy security improvement (European Commission, EC 2013,EC 2014 b, c, d, e, f)
- Reduction of GHG emissions
- Maintaining sustainability

Energy intensity is defined as the ratio between the gross input consumption of energy usage and the GDP per year. Energy intensity is the important economic indicator of energy consumption, including overall efficiency. Energy intensity is expressed in MTOE expressed in currency of the respective countries: dollars for the United States, rupees for India. An industry with higher production with higher mix will have more energy intensity than energy intensity for service sectors, even if energy efficiencies of them are identical. A nation that is dependent on imports of carbon-intensive goods have lower intensity than the countries that manufacture goods for exports when all other factors are identical. Energy intensity is final consumption percentage of GDP.

Table 1.1 provides overall information of world best practice energy intensity of key processes in primary iron and steel production blast furnace with BOFs. The total final energy, including thermal and electrical in raw material preparation like coke production and sintering, is 2.7 GJ/ton. The final energy consumption for iron-making process is 12.2 GJ/ton, and steel-making process with BOFs is 0.22 GJ/ton. The total overall final energy consumption, including thermal and electrical for complete hot rolled strip plants, is 16.7 GJ/ton.

Energy intensity of GDP is a vital parameter for economic growth and sustainability. Energy intensity on higher part indicates higher price of cost of energy

TABLE 1.1
World Best Practice Energy Intensity of Key Processes in Primary Iron and Steel Production (BF-BOF Route)

Process	Thermal Energy [GJ/ton]	Electricity [GJ/ton]	Final [GJ/ton]
Material preparation			
Coke production (coke oven) European Union (EU) [28]	0.7	0.1	0.8
Sintering	1.7	0.2	1.9
Iron-making process			
Iron ore reduction (blast furnace)	12.1	0.1	12.2
Steel-making processes			
Hot metal oxidation (BOFs)	0.12	0.1	0.22
Continuous cast	0.05	0.05	0.1
Hot rolling—strip	1.3	0.3	1.6
Hot rolling—bar	1.6	0.3	1.9
Hot rolling—wire	1.7	0.4	2.1
Cold rolling	0.2	0.3	0.5
Total (based on hot rolled strip)	15.9	0.8	16.7

Source: World Steel Association report https://worldsteel.org/

conversion into GDP. Energy intensity on lower part indicates low price of cost of energy conversion into GDP. Natural disaster and energy calamities disruption of services due to natural services and stochastic economic shocks like wars and massive power outages highly affect energy intensity of respective nations. Geographic locations and climatic conditions also vary for energy efficiency and its use. Energy intensity is an important factor for per capita GDP, and acts as a reference indicator for growth of a country and its development. Different nations have higher or lower energy intensity. The underdeveloped countries like Bangladesh have high energy intensity and therefore low standard of living. Energy intensity of Russia is just double than that of the United States and it is also dependent on the climatic conditions and terrestrial space. Italy is one of the most developed country and the most energy-efficient country in industrial global world. But due to the COVID-19 pandemic affecting most developed and developing countries, the performance indicators like energy intensity and GDP are inversely proportional and there are sudden transients in the global economy with the pandemic and prevailing unavoidable circumstances. Energy efficiency, conservation and sustainable development is explained in detail in Chapter 3 and in Chapter 8, with regional and global perspective with special emphasis on the COVID-19 pandemic. India has better energy intensity as compared to Japan and China. The denominator term, that is GDP, is an important factor to compare and study the economic growth rate between the countries for developing and developed countries. In a comparative study of nations for estimating energy intensity, the most important indicator is GDP conversion, which is denominator of energy intensity. By GDP variation up and down, energy intensity will vary— and therefore, economic development of the country will vary according to GDP. The detail of financial indicators related to project implementation and execution is explained in Chapter 6 with respect to project cost and financial analyses, including risk analysis. The Organization for Economic Cooperation and Development World Energy Outlook 2013, 2014 OECD/IEA, Paris. IEA (2013, 2014) (OECD) mainly focused on the following four valuable points for comparison of GDP.

1. Currency should be identical.
2. Methodology and measurement should be identical.
3. Definition of GDP should be same.
4. Price level should be same.

1.5 ENERGY EFFICIENCY AND CONSERVATION ACTIVITIES: METAL INDUSTRIES

The major consumers of energy are metal industries including copper, aluminum and iron and steel industries. Energy utilization, sustainability and security are key performance parameters for studies. The author believes that a global informatics center for energy conservation and efficiency should be developed in which all archiving of data for 200 years would be available and accessible across the cloud. Energy management in energy-intensive industries such as iron and steel industry and foundries is a very important and main focus area of energy conservation in metal industries. ISO 50001: Energy Management System (2011 and 2018) implemented a new program and standards. To correctly implement an energy management system, the first step is to conduct an energy survey audit that examines the ways energy is currently used in the utilities, and identifying some measures for reducing energy costs. Energy efficiency and conservation activities with special reference to metal industry are discussed in detail in Chapters 3 and 4.

Figure 1.1 provides overall production of steel for the year 2017. China is a major contributor of steel production of 831.7 MT. The second-largest producer of steel is Japan, producing 104.7 MT, and India is the third-largest producer, at 101.4 MT. The United States produces 81.6 MT of steel, Russia 71.3 MT, South Korea 71.1 MT, Germany 43.6 MT and Italy—24 MT.

Figure 1.2 gives overview of the primary and secondary steel production process. It provides schematic representation of primary and secondary steel-making processes, with four different classification and subsection like raw material preparation, iron-making process, steel-making process and finishing. There are two design schematics for the production of steel. Primary steel is produced by optimum utilization of iron ores and scrap, and production of secondary steel using sponge iron and scraps. There are various steel products in the global market from slabs and ingots to thin sheets in which large volume of manufacturing industries are involved (ACER/CEER, 2014).

The input raw material for producing primary steel is pig iron, which is produced in blast furnaces using iron ore, dolomite, coke and (injected) coal as raw materials, which is as energy intensive as production of coke from coal in a coke-oven plant. The primary steel is produced by reduction of iron ore and it is a high energy–consuming process. It also produces blast furnace gas, which is used for heating (similar to

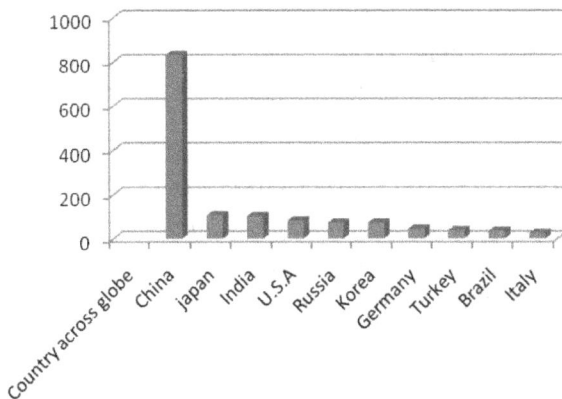

FIGURE 1.1 Major steel-producing countries in the world[1] (2017 production in million tons [MT]). (https://worldsteel.org/media-centre/press-releases/2018/world-crude-steel-output-increases-by-5-3-in-2017)

1 World steel Association report. https://worldsteel.org/

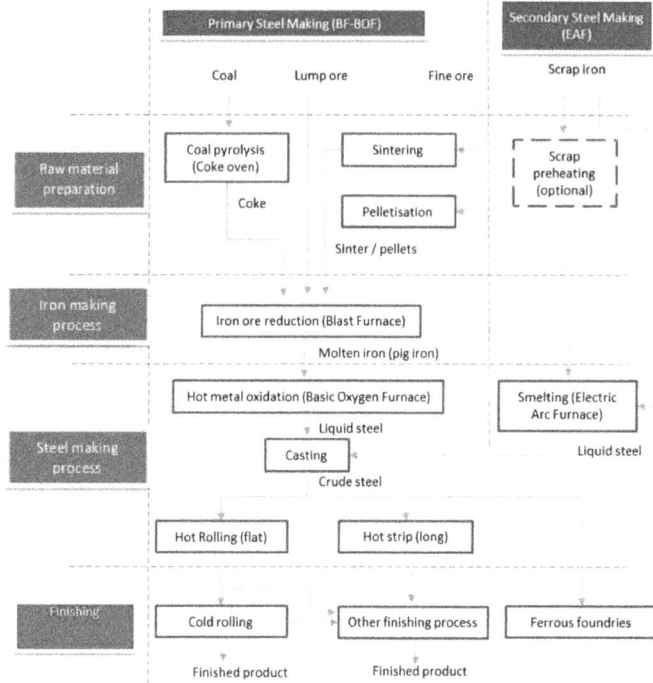

FIGURE 1.2 Outline of the production process of primary and secondary steel processes.[2] Brussels (European Commission, 2014a)

coke oven gas) and also, electricity if top gas pressure recovery turbines are installed and slag, which is used as a building material. Steel is produced by BOFs and open hearth furnaces (OHFs). BOFs are more rapid in comparison with OHFs because of its high productivity and lower capital cost. In BOFs, the steel quality can further be improved by ladle refining processes used in the steel mill. Electric arc furnaces are used for melting and refining of sponge iron and scrap. DRI is produced by reduction of ores below melting point with different properties.

1.5.1 ENERGY EFFICIENCY, CONSERVATION AND PROCESSES OF STEEL PLANTS FROM AN INDIAN PERSPECTIVE

Metal industries, starting from metal's invention and discovery, contribute to be a very important part of humankind and in our day-to-day lives. India plays a significant role as the third-largest producer of steel, and it will be second-largest producer of steel in coming days. The sector contributes to 2% of the country's GDP. It has been identified as one of the most energy-intensive sectors and it (if annual energy consumption is more than 30000

2 Steel making process layout for European Union and India. Brussels. EC (European Commission) (2014a)

toe) is certified as designated consumers by the Bureau of Energy Efficiency in India. The complex process, along with several utilities, paves the way for tremendous energy efficiency potential in the metal industry. The metal sector is inclusive to large-, medium- and small-scale industries. The efficiency in some plants are at par with the world's best plants. However, a good number of plants still have significant reduction potential.

Figure 1.3 gives overall methodology which is followed for Performance Achievement Trade (PAT) Cycle-1 in India (PAT BEE-2018) to achieve energy-saving targets. It includes data collection, data analysis and feedback from PAT stakeholders and success stories, including case studies which are to be shared with plant team. Data collection includes sector-specific data from the Bureau of Energy Efficiency, GDP as per the Indian perspective, benchmarking and miscellaneous data for secondary research. Data analysis includes in-depth studies according to PAT scheme in specific energy consumption and impact of energy intensity. The next steps are the feedback from stockholders who are key personnel involved sector-wise in industrial processes, and contacting designated consumers.

Figure 1.4 shows overall energy savings achieved in the PAT scheme by the iron and steel sector as a designated consumer. Energy savings in terms of oil and coal

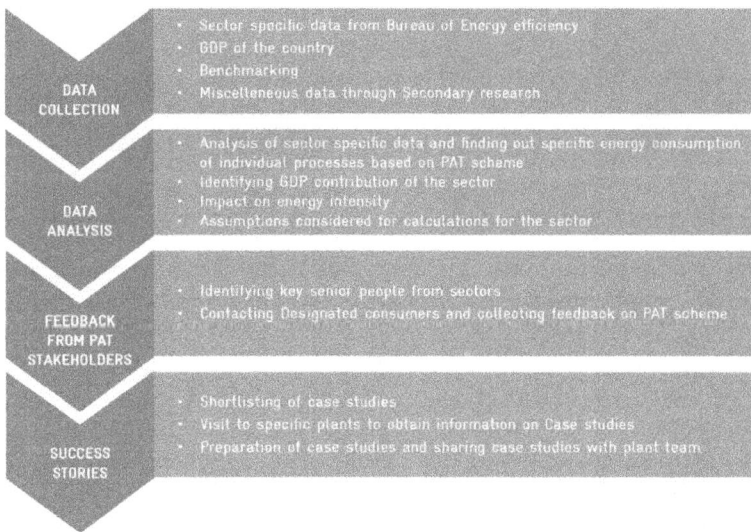

FIGURE 1.3 Methodology followed for impact assessment of PAT Cycle-1 in India.

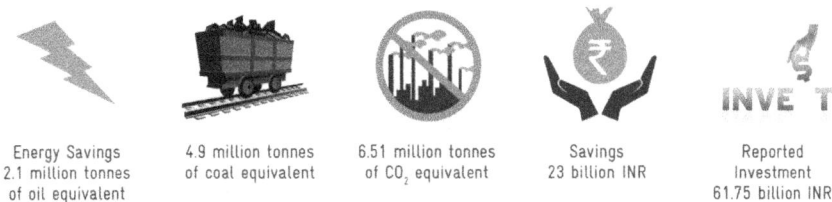

| Energy Savings 2.1 million tonnes of oil equivalent | 4.9 million tonnes of coal equivalent | 6.51 million tonnes of CO_2 equivalent | Savings 23 billion INR | Reported Investment 61.75 billion INR |

FIGURE 1.4 Savings achieved by the iron and steel sector in PAT. https://beeindia.gov.in/sites/default/files/press_releases/Consolidated%20Report.pdf [6]

TABLE 1.2
Reduction in CO_2 Emissions from PAT Cycle-I

Parameter	Value
Reduction of CO_2 emissions due to implementation of PAT Cycle-I in all sectors	31 million tons of CO2 equivalent
Reduction of CO_2 emissions due to implementation of PAT Cycle-I in iron and steel sector	6.51 million tons of CO2 equivalent
Contribution to CO_2 emissions reduction in overall PAT Cycle-I in iron and steel sector	21%

equivalent is 2.1 million tons and 4.9 million tons, respectively, with emission reduction of 6.51 million tons of CO_2 equivalent, and cost savings reported is 23 billion INR. Reported investments, to achieve this, was INR 61.75 billion.

Table 1.2 shows reduction in CO_2 emissions from performance achievement and trade cycle in phase-1 cycle of India in which significant reduction of CO_2 emissions are reported. Reduction of CO_2 emissions due to implementation of performance achievement and trade Cycle-I for all sectors was 31 million tons of CO_2 equivalent and 6.51 million tons of CO_2 equivalent in the iron and steel sector; overall contribution of iron and steel sector to PAT Cycle-I is 21% of total.

1.5.2 DISCUSSION AND ANALYSIS

The present is a difficult time due to the COVID-19 pandemic, which results in demand and supply mismatches. The oversupply and low demand was seen in the oil and gas industry, and more specifically on the demand side. Due to the pandemic, the cost of oil fell from 60 USD per barrel at the beginning of the year to 15 USD per barrel due to the pandemic [7]. The estimates for current demand dropped compared with 2019, ranging between 65 million barrels per day and 80 million barrels per day, which results in excess production of 20–35 million barrels per day. Because of the drop of consumption and no check on supply, the excess production has been resulted to storage. (COM (2014) 654 final, October, Brussels. EC (2014m). The storage capacity is very closed to saturation point, and Saudi Arabia is one of the key leaders of oil global storage capacity. (Communication from the Commission, Europe 2020) (COM 2010, COM 2011). The current impact of COVID-19 is not limited to the oil and gas market, but to global society as a whole.

The overall economic global impact from the crises will also fundamentally alter the demand for oil and gas, and will take substantial time to normalize, and companies will work to bring supply chains closer to normal to avoid future dips. The restoration of business models which are corrective actions for the oil and gas industry, and also other industries like the metal industry—to reduce head counts, close nonperforming businesses, and close and merge field offices—will reduce operational expenditures. The COVID-19 pandemic crisis will be overcome by accelerating initiatives to automate work processes wherever possible to reduce and minimize manpower, and with the implementation of artificial intelligence, visual reality and robotics, the progress in work for process industry like oil and gas, metals, etc., will be accelerated. Integrating operational data with supply chains and using analytical tools will not only help to evaluate the bottlenecks and their impact on operations, but will make supply chain more efficient.

The report of intergovernmental panel on climate change IPCC of 2018 targeted that temperature of global warming should be limited to 1.5°C by 2050 [8]. The iron and steel industry plays a significant role and is responsible for 7% of all anthropogenic CO_2 emissions (JFE Steel CSR Report for 2016).

As per IPCC report and studies, carbon emission should be significantly cut and tCO_2/t steel should be reduced. High available innovative technologies in all processes will decrease emissions by 15–20%. The novel technologies which are adopted for reducing CO_2 emissions is by CO_2 capture and storage (CCS). CCS technology is most applicable in the oil and gas industry and the iron and steel industry. Numerous research initiatives have been launched for hydrogen production and reduction of iron oxides. The future trends are green electricity and hydrogen, which will be linked together. With the complete modernization of production, energy utilities and adopting new innovative technologies, the "world steel" industry could reach the minimum level of carbon emission of 0.4–0.5 tCO_2/t steel and thus reducing 65% of current annual emissions.

1.6 CONCLUSION

In this chapter, historical perspective with special reference to the iron and steel industry and energy scenarios are discussed. Energy conservation and its opportunities—with relevant examples and case studies—are studied. The performed study in this chapter provide a basic fundamental idea of potential and future application of iron and steel making and its byproducts by highlighting both positive and critical aspects in close coordination with other sectors.

Focusing on the iron and steel industry, new technological solutions need to be developed to achieve quality byproducts in order to increase their utility for sustainability and to reduce environmental impact (COM 2010) (EEA 2014). It is a pathway to achieve the "zero waste" goal in the steel sector and conservation measures to some natural resources, and reduces the environmental impact of the production (European Union Greenhouse Gas Inventory 1990–2012 and Inventory Report 2014). Implementing the concept of circular economy—that is: replace, reuse, recycle and restore—will provides energy-saving measures and to achieve higher efficiency in the production processes. The future trends of the iron and steel industry are not only in innovation and new technological solutions, but also reducing negative environmental impacts to make processes cleaner and more environmentally friendly.

REFERENCES

ACER/CEER (2014); Agency for the Cooperation of Energy Regulators/Council of European Energy Regulators.

Brussels EC (2010) (March); Communication from the Commission, Europe 2020, A Strategy for Smart, Sustainable and Inclusive Growth, COM(2010) 2020 final.

Brussels EC (2011a); Communication from the Commission, Delivering the Internal Electricity Market and Making the Most of Public Intervention, COM(2013) 7243 final.

Brussels EC (2011b); Communication from the Commission, the European Union Energy Policy: Engaging with Partners Beyond our Borders, COM(2011) 539.

Brussels EC (2013a) (January); Communication from the Commission, Energy Prices and Costs in Europe, COM (2014)21/2, Report from the European Commission, Annual Growth Survey 2014.

Brussels EC (2013b); A Single Market for Growth and Jobs: An Analysis of Progress Made and Remaining Obstacles in the Member States, COM (2013) 785 final.

Brussels EC (2013c); European Union Energy, Transport and GHG Emissions Trends to 2050: Reference Scenario 2013.

Brussels EC (2014a); Communication from the Commission, European Energy Security Strategy, COM (2014) 330 final.

Brussels EC (2014b); Autumn 2014 Forecast, DG Ecfin.

Brussels EC (2014c); Energy in Figures, DG Energy.

Brussels EC (2014d); Benchmarking Smart Metering Deployment in the EU-27 with a Focus on Electricity, COM (2014).

Brussels EC (2014e);

Brussels EC (2014f); The Consumer Scoreboard.

Brussels EC (2014g); Communication from the Commission: Guidelines on State Aid for Environmental Protection and Energy 2014–2020, OJ C 200, 2014, 6 28.

Brussels EC (2014h) (January); Communication from the Commission, A Policy Framework for Climate and Energy in the Period from 2020 to 2030, COM (2014) 015 final.

Brussels EC (2014i); Communication from the Commission, Energy Efficiency and Its Contribution to Energy Security and the 2030 Framework for Climate and Energy Policy, COM(2014) 520 final.

Brussels EC (2014j); Impact Assessment Accompanying the Document Communication from the Commission to the European Parliament and the Council, Energy Efficiency and Its Contribution to Energy Security and the 2030 Framework for Climate and Energy Policy, SWD(2014) 255 final.

Brussels EC (2014k); Impact Assessment Accompanying the Document Communication from the Commission to the European Parliament, the Council, the European Economic and Social Committee and the Committee of the Regions A Policy Framework for Climate and Energy in the Period from 2020 Up to 2030, SWD (2014) 015 final.

Brussels EC (2014l) (October); Communication from the Commission on the Short Term Resilience of the European Gas System: Preparedness for a Possible Disruption of Supplies from the East During the Fall and Winter of 2014/2015, COM (2014) 654 final.

Brussels EEA (European Environmental Agency) (2014) (November); Communication from the Commission, Energy 2020, A Strategy for Competitive, Sustainable and Secure Energy, COM (2010) 639 final.

Bureau of Energy Efficiency (India) (2018) (September); Performance Achieve and Trade Iron and Steel Sector Report "Ministry of Power in Partnership with GIZ, Germany".

Cabinet Office (2005); World Economic Trend, Spring.

EC (European Commission) (2014); Brussels 2013, Annual Report on the Results of Monitoring the Internal Electricity and Natural Gas Markets.

Ernest Orlando (2008); World Best Practice Energy Intensity Values for Selected Industrial Sectors, Lawrence Berkeley National Laboratory 2008.

European Council (2014); Annual European Union Greenhouse Gas Inventory 1990–2012 and Inventory Report 2014, Submission to the UNFCCC Secretariat, Copenhagen, European Economy Statistical Annex (June), DG Ecfin.

Eurostat & OECD (2005); Purchasing Power Parities and Real Expenditures: 2002 Benchmark Year, 2004 Edition.

IEA (International Energy Agency)l (2014a); Conclusions on 2030 Climate and Energy Policy Framework, SN 79/14; http://www.consilium.europa.eu/uedocs/cms_data/docs/press-data/en/ec/145356.pdf (accessed 23 October 2014).

IEA (International Energy Agency) (2014b) (2013); World Energy Outlook 2014, OECD/IEA, Paris.

IEA (International Energy Agency) (2014c); Energy Statistics of Non-OECD Countries 2014, OECD/IEA, Paris.

IEA (International Energy Agency) (2014d); Energy Balances of OECD Countries, OECD/IEA, Paris.

IEA (International Energy Agency); "World Development Indicators" World Bank "Handbook of Energy & Economic Statistics in Japan" Institute of Energy Economics, Japan.

JFE Steel CSR Report (2016); http://www.jfe-holdings.co.jp/en/csr/pdf/csr2017e.pdf.

John Williamson (1983); The Open Economy and the World Economy.

Ministry of Steel Annual Report (2017–18); Government of India.

MoC (2018); Annual Report 2018; Department of Industrial Policy and Promotion, Ministry of Commerce and Industry; http://dipp.nic.in/sites/default/files/annualReport_English_08March2018.pdf.

Noriki Hirose (2005); ESRI Discussion Paper No. 153, Cabinet Office Energy Balances of OECD/Non-OECD Countries.

World Bank (2005); Purchasing Power Parities: Statistics to Describe the World. World Bank.

World Bank (2014); World Bank Data; http://data.worldbank.org/ (accessed 1 September 2014).

World Energy Outlook (2013); OECD/IEA, Paris.

World Steel Association Report; www.worldsteel.org.

WEBSITES

[1]   https://www.britannica.com//.
[2]   https://www.worldenergy.org
[3]   https://www.energy.gov/eere/amo/articles/advanced-manufacturing-office-update-january-2015
[4]   https://www.energy.gov/eere/amo/articles/advanced-manufacturing-office-update-january-2016
[5]   www.bee-india.nic.in
[6]   https://beeindia.gov.in/sites/default/files/press_releases/Consolidated%20Report.pdf
[7]   https://www.nytimes.com/2020/04/20/business/oil-prices.html
[8]   https://www.ipcc.ch/sr15/

2 EE&C Policy Considerations

Binoy Krishna Choudhury

2.1 ENERGY, ENVIRONMENT, ECONOMY AND SOCIETY

2.1.1 UNITED WE PROGRESS, DIVIDED WE REGRESS

Case Study 2.1: US Steel Conglomerate Nucor

In the year 1965, when noted metallurgist-cum-industrialist Ken Iverson was asked to become the President of US Steel conglomerate, Nucor, its only profitable division was the steel girder–making operation looked after by him. Following 16 years as president of Nucor, Iverson consistently moved on with restructuring the administration and breaking the old hierarchical modes of operation; rather emphasizing teamwork, encouraging performance with compensation, sharing interest, performance improvement of mini-mills and community involvement. His continuous innovation in management, technology (including energy efficiency and productivity) and policy framework started yielding results within two years when the company turned around to become profitable as a whole and subsequently in 14–16 years helped the then-declining US steel industry to grow again (World Steel Association, 2012). This success story demonstrated coherent social, energy and economic policy initiatives and their consistent application that had impacts beyond the company, to the economy of the country.

Case Study 2.2: Iceland's Transition from Fossil Fuel–Based to 100% Renewable Energy–Driven Economy

Now, from a company to a country—let us take the case of Iceland's transition from fossil fuel–based developing economy in the 1970s to the present developed economy propelled by 100% renewable energy–generated electricity. Iceland, having hydro, geothermal and wind power among renewable energy resources in a small and peaceful state, in the 1970s, was emerging out of centuries of foreign rule and poverty. The weak economy of the country was not able to absorb the huge fluctuations of international oil prices. Other barriers included lack of knowledge of renewable energy sites and potential, absence of institutional frameworks for critical finances and relevant skills, technology and manpower. Moreover, the energy mix—price, balance of supply and demand, and

---

DOI: 10.1201/9781003157137-2

use pattern—of a country in general is a complex equation of multiple variables including politics, culture and practice. However, Iceland faced that challenge with determined and cohesive local municipalities, government and people at large, and was able to demonstrate strong teamwork in a coherent policy framework. Thus, in an astounding speed, Iceland developed short-, medium- and long-term plans, and implemented them, to assess the potential, mobilize finances and deploy manpower with the help of innovative schemes. For example, unsuccessful drilling costs for geothermal energy capture were paid by the government to absorb losses with the help of profits for successful ventures. Now, Iceland is declared a developed nation, ranked fourth in terms of human development index with value of 0.949 in 2020 (UNDP, 2020), and is providing technology and support to more than 50 countries with the geothermal energy exploration and utilization technologies, including the world's largest plant being installed in China (UNDP, 2021). Many countries in the world have sufficient renewable energy sources to meet all their needs but lack policy implementation to help achieve such targets, in a scenario of similar challenges.

Policy has the power to even overcome the resource constraints of any country and could be a win–win situation for efficient use of energy, improved environment and economy in a modern society. Elbaum (2007) clearly showed the advent of the iron and steel industry during 1900–1973 in Japan in spite of huge deficiency in domestic availability of coke-grade coal and iron ore. Even during a crisis, policy needs to be coherent even at the short term. During World War II, even public gate iron bars were cut and recycled to meet the wartime need, as shown in Figure 2.1.

FIGURE 2.1 Remnants of iron fence bars in York Whip-Ma-Whop-Ma-Gate. Such public property fences were sawed for the iron and recycled during World War II [1].

Case Study 2.3: Metal Industry Planning and Policy Preparation of the European Union

Similarly, cases are interesting for the continents, too; for example, the European Union. Europe was the cradle of Industrial Revolution (1760–1840) and two world wars during 1914–1918 and 1939–1945. The former brought mechanization of modern civilization and subsequently also global warming and industrial pollution. The latter brought much death, hunger and collapse of economies—and subsequently, associations for peace and prosperity, viz. the United Nations (1945) and European Union (1993). However, the seeding of the European Union was started in 1950 when the European coal and steel communities began to unite for peace and prosperity. This led to an evolution process—the European Economic Community, formed in 1957, gradually evolved into the European Union in 1993 ensuring "four freedoms"; viz. the 27 member countries' movement of goods, services, people and money. The union has a total area of over 4 million km^2 and an estimated total population of about 447.7 million (European Commission, [a]). The steady and proactive policy approach of European Union has led to a happier Europe, secured a Nobel Peace Prize in 2012, overcame many challenges including the COVID-19 pandemic and set the stage for the rare opportunity for Europe to become the first carbon-neutral continent in 2050. Interestingly, since the beginning in 1950 until now, the steel community—presently, notably the European Steel Association (EROFER)—has taken a prominent role in action planning and policy preparation of the European Union.

A noticeable evolution in policy imperatives of metal industries is the gradual inclusion of broader perspectives as the prevalent challenges were overcome. Since the early stages of revolution of the industry, mining of ore and its processing technologies have changed toward improvement of productivity and profitability. In the last few decades, the focus has been shifting toward inclusive growth, environmentalism and sustainability. Interestingly, energy efficiency and conservation (EE&C) policy has always been crucial in this journey, providing breakthrough lines of action for environment, economy and society.

2.1.2 SOCIAL EVOLUTION WITH ENERGY, ENVIRONMENT AND ECONOMY

As we deal with any social issue nowadays, the roles of energy, environment and economy have become so inevitable that their governance has become a standard practice in metal industries, as well. These issues are aptly reported as environmental social and governance (ESG) reporting, which is discussed in Chapter 8. Modern tools help the planners to deal with multiple criteria in the decision-making process of arriving at the optimal solution to fulfill the corporate objectives. This is particularly important for metals and mining industries, which have been reported to have the highest ESG risk score of 11 with compared to risk score of 6 for the engineering and construction industries (S&P Global, 2019).

2.2 EE&C AND SUSTAINABLE DEVELOPMENT

2.2.1 ENERGY TRILEMMA

In 2015, the World Energy Council observed that Paris Agreement targets may remain commitments far from achievements if the three pillars—viz. energy security, energy equity and environmental sustainability—are not equally strong to meet the 17 Sustainable Development Goals established by the United Nations, which are all linked to efficient use of energy sources and its conservation (World Energy Council, 2015). The report points out that the policy should be flexible to accommodate the regional goals to remain strict at meeting the global objectives. A number of toolkits, including the World Energy Trilemma Index, are available for individual or collective use. Policy needs to be holistic to also help public and private finance flow for long-term high return on investment rather than short-term attractions.

Following are a few examples showing the immense importance of policy framework (say, for the metal industry) at local, national and regional or international levels.

1. At the user and industry level: It is possible to reduce CO_2 emissions on average by 200 tons per year by the judicious use of every one ton of copper in the energy sector (European Copper Institute, 2013). However, for production of one ton of primary copper, up to nine tons of CO_2 eq emission is possible (Nilsson et al., 2017). It may be noted that this judicious use will include identification of opportunity, application of appropriate technology, monitoring throughout the life cycle and recycling at appropriate time, which involve many stakeholders including government, industry, consultants and service providers, etc.

2. At the country level: The iron and steel industry in India is one of the core industries. India is the third-largest steel producer in the world. India's crude steel output in 2019 was 111.4 million tons per annum, contributing to 6% of the world's crude steel production (Steel Statistical Yearbook 2020 Concise Version, World Steel Association, 2020; www.worldsteel.org/steel-by-topic/statistics/steel-statistical-yearbook.html). India Energy Security Scenarios 2047 (IESS, 2047) is an energy scenario building tool which aims to explore a range of potential future energy scenarios for India, for diverse energy demand and supply sectors, leading up to 2047, i.e. 100 years after India's independence in 1947. The open platform (http://iess2047.gov.in/) has provision for updating 21 energy demand and supply sectors, and 50 levers that will affect India's energy system and are available to the user. Combinations of these choices offer hundreds of energy pathways until 2047. It has provision for users to modify each sectors and levers to create a scenario or select any one from the 11 given scenarios including the least effort scenario (LES, all at level 1), maximum renewable energy pathway (MREP), default, etc. The projected import percentage for coal, oil, gas and overall in 2047 for LES are 87%, 95%, 69% and 84% respectively, whereas for MREP, they are 13%, 77%, 26% and 23%. Thus, because of the change of scenario from LES to MREP, coal import percentage of total consumption in India may change from as high as 87% to as low as 13% in 2047.

3. At the regional level: Asia Pacific Economic Cooperation (AEPC) was formed in 1989 to foster cooperation in the region for sustainable and inclusive growth among member economies, currently 21 countries: Australia; Brunei Darussalam; Canada; Chile; Chinese Taipei; Hong Kong, China; Indonesia; Japan; Malaysia; Mexico; New Zealand; Papua New Guinea; People's Republic of China; Peru; Republic of Korea; Singapore; Thailand; The Philippines; the Russian Federation; the United States; and Vietnam. Thus, five of the world's top ten producers belong to this region with a share of more than 85% of steel produced (Table 2.1). In a report published in 2013, much emphasis was made on realization of huge energy-saving potential from mandatory efficiency standards. While recommending on setting up a national strategic framework for developing roadmaps, further follow up and implementation has not been reported on EE&C drive. APEC member economies achieved commendable success as they developed a comprehensive ten-year strategy in support of the region's routine vaccination effort, also to enhance the resilience and sustainability of immunization programs in the region through 2030. APEC as an institution, through good governance and stakeholder engagements, has committed to continuous improvement through advancement of the APEC Putrajaya Vision 2040 with a spirit of equal partnership, shared responsibility, mutual respect, common interests and common benefits. An appropriate implementation plan and review of its progress is lacking at the present moment, but is expected in the future if APEC is to utilize its huge potential and maintain its unique position as the premier forum for regional economic cooperation, as well as a modern, efficient and effective incubator of ideas including in the area of EE&C [2]. Another world regional body, the European Union, worked out a relevant "Steel Action Plan" to identify and resolve problems associated with the steel industry (European Commission, [b]), as also discussed in Section 2.3.

4. At the global level: An ideal example is United Nations Framework Convention for Climate Change (UNFCCC) established in 1992 for supporting and acting on global response to the threat of climate change and has nearly universal membership of 197 countries. The objectives are to meet the aims of the Kyoto Protocol, and more recently those of the Paris Agreement, to keep the global average temperature rise this century as close as possible to 1.5°C above pre-industrial levels, and in a longer run, to stabilize greenhouse gas (GHG) concentrations in the atmosphere at a level that will keep dangerous human interference with the climate system under complete control within a timeframe to allow ecosystems to adapt naturally and enable sustainable development. Energy efficiency and renewable energy are two of the most important tools to combat the climate crisis as energy sector has been identified to be the largest contributor to GHG. Race to Zero, backed by the UN, was launched in 2019 toward achieving (medium-term objective) halving (50%) the emissions compared to pre-industry levels by 2030 and ultimately (long-term objective) net zero carbon emissions by 2050 at the latest. This largest-ever alliance of "real economy" actors from 120 countries now mobilizes a coalition of leading net zero

initiatives, representing 733 cities, 31 regions, 3,067 businesses, 173 of the biggest investors and 622 higher education institutions collectively now cover nearly 25% of global CO_2 emissions and over 50% GDP of the world economy. The aim (short-term objective) is to gain the necessary momentum from the Glasgow Climate Conference (COP 26) in November 2021. Out of many long term policy decisions taken in Glasgow, the one expected to drive metal industries to a new horizon is to phase down unabated coal power and inefficient subsidies for fossil fuels. Another initiative by UNFCC in 2019 was signing up partnership with the International Renewable Energy Association (IRENA), the global intergovernmental organization with 159 member states, and the European Union. IRENA has supported members/ entities to achieve an 100% renewable energy–driven economy [10].

2.2.2 Challenges before Metal Industries

Challenges for sustenance of metal industries are location specific; however, a generic approach is followed in this chapter for clarity and usefulness of the readers.
Main challenges of metal industries are as follows:

- Availability and cost of raw materials.
- Availability and cost of energy resources.
- Regulations associated with environmental and climate change.
- Competition from the same metal industry within the country or from other country.
- Competition from substitute metal industry; for example, steel is substituted by aluminum alloy in transport industry, or vice versa.
- Lack of coherent policy framework so that the overall objective fulfillment is not jeopardized because of inconsistency among industry policy, trade policy, FDI policy, EE&C policy, etc.

Another challenge posed to metal industries which may be noteworthy here is the COVID-19 pandemic situation, which suddenly reduced market demand and hampered entire supply chains, particularly during lockdown, for example price fluctuations and availability of raw materials. In the United States, steel production dropped to 72.7 million tons in 2020 from 87.8 million tons in 2019, i.e. a reduction of 17.2% in just one year (Table 2.1)

2.2.3 Overcoming Challenges before Metal Industries

Some of the common challenges before metal industries (European Commission, [c]) may be overcome by appropriate action plan including the following:

- Make the regulatory framework right. This may be a never-ending process of continual improvement as explained in a plan–do–check–act (P-D-C-A) cycle. It may always strive to provide a level playing field for the metal industry, helping the internal market function properly, for investor certainty and predictability for sustainable development.

TABLE 2.1

Top Ten Steel-Producing Countries in 2020 and Their Emissions, Population, Human Development Index (HDI) and Rank among the Top 10 (Absolute and per Capita Value-Wise)

World Rank	Country	Steel Production (in Million Tons/Year)			Population	Per Capita Production		CO₂ Emission (Mt and Ton/Year/Head)				Human Development Index (HDI)	
		2020 (Mt)	2019 (Mt)	% Diff.	2015 (M)	kg/yr/head	Rank in 10	2018	Rank in 10	Per Capita	Rank in 10	2020	Rank
1	China	1053	1,001	5	1,371	730	3	10,313	1	7.5	7	0.76	9
2	India	100	111	−11	1,310	85	10	2,435	3	1.9	10	0.65	10
3	Japan	83	99	−16	127	781	2	1,106	5	8.7	4	0.92	3
4	Russia	73	72	3	144	497	4	1,608	4	11.2	3	0.82	5
5	United States	73	88	−17	321	274	8	4,981	2	15.5	1	0.93	2
6	South Korea	67	71	−6	51	1400	1	631	7	12.4	2	0.92	4
7	Turkey	36	34	6	79	429	6	413	10	5.3	8	0.82	6
8	Germany	36	40	−10	82	485	5	710	6	8.7	5	0.95	1
9	Brazil	31	33	−5	204	159	9	428	9	2.1	9	0.77	8
10	Iran (e)	29	26	13	78	326	7	629	8	8.0	6	0.78	7

Source: PRESS RELEASE—Global crude steel output 2019 thru 2020; www.worldsteel.org, http://hdr.undp.org/en/content/latest-human-development-index-ranking and https://databank.worldbank.org/ accessed on 1 March 2022 [9]

- Strive toward 100% recycling and zero waste of metals for multiple benefits to the society, viz. more than 90% saving of energy and resources, minimize unnecessary movements of goods and services, circular economy benefits to a larger number of people, reducing pollution and saving our environment toward sustainability.
- Enhance the market demand of metals. This can be accomplished by studying the market of the metals and promoting the subsector to enhance the demand of that metal. For example, steel demand may be enhanced by promoting the building and automobile sectors. All growth-oriented initiatives may be considered.
- International negotiations for accessing each other's market at a level plane. Often the countries impose restrictions on import/export of ores/metals/metal products in favor of the interest of the domestic industry. Sometimes, negotiations can open up the opportunities for win–win situations of both the countries in favor of sustainable development of the concerned metal industry on behalf of international bodies, such as the United Nations, European Union and international associations.
- Facilitate access to the raw materials. One country may have all other conditions favorable except the availability of sufficient raw materials. In such cases, short-, medium- and long-term strategies can help ensure the metal industry access to the raw materials from foreign lands or new explorations undertaken within or outside the country home to the metal industry. For example, Japan is able to maintain an assured supply of raw materials from other countries and produce quality steel exported all over the world. The European innovation partnership on raw materials is another such example.
- Trade is particularly important for metal industries, which are normally far away from the raw materials source and the market. A free trade agreement (FTA) or negotiations at bilateral level to support liberalization of international trade under the World Trade Organization (WTO) would benefit all stakeholders.
- Energy costs represent about 20–70% of total production costs, varying from metal to metal, availability of hydropower, tariff policies, etc. Industries face more challenges in some regions having higher energy prices further amplified recently. In short and medium terms, a shortfall in production due to high energy costs may be overcome by enhancing compensatory production more from recycled metals, as secondary metal manufacturing processes may take only 5% of the energy consumed in primary metal manufacturing processes. A strategic import policy of recycled metals may come out to be a savior in some critical condition.
- Research and innovation is a continuous process critical for sustenance of metal industries. Green manufacturing of metals, using hydrogen and electricity as the main source of energy, looking for new products in niche markets such as super-conductors, reaping the benefit of circular economy in the entire life cycle of metals, and involving all the stakeholders and their participation in the contributory role both in the industry and the society are the obvious means of overcoming the challenges therein. Notably

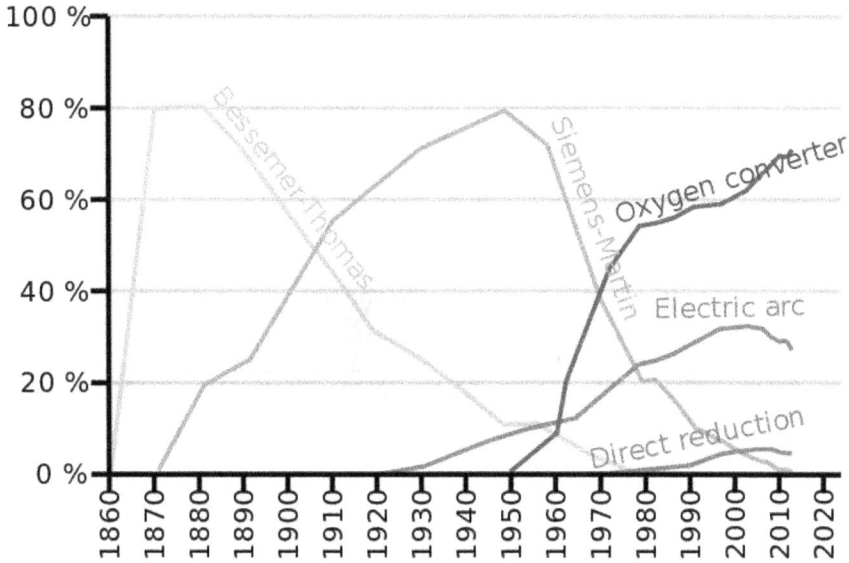

FIGURE 2.2 Distribution of world steel production by various production methods evolution over time.

Source: https://upload.wikimedia.org/wikipedia/commons/2/23/Evolution_convertisseurs.svg

among metal industries, iron and steel production technology has been continuously evolving with time, to cater to the need of the hour. Now the focus is on EE&C and transition from gray to green technology, whereas in the past, it was productivity and convenience. Distribution of world steel production by various production methods evolution over time is shown in Figure 2.2.

- Enhancing flexibility, product quality and capital inflow by ensuring availability of raw materials and energy, as well as proximity to end-users, downstream industries are important to overcome the challenge of shifting the existing metal industry to countries with lower energy prices and lower social and environmental costs.

- Continuing business in view of the non-renewable nature of natural resources, particularly the ores, is a challenge to be overcome through long-term planning. Reserves of a metal are estimated from the technically and economically viable explorable ores in a country or region or globe. This quantity, when divided by production, gives us a ratio called reserve to production ratio and is generally expressed in number of years that ore is expected to be available for the industry, assuming these two quantities are based on current estimates. Actually, however, the reserve may increase if the new discovery of exploitable resources outpaces the rate of consumption. Annual global production and reserves data for selected metals during the period from 1956–2018 and their ratios are shown in Figure 2.3.

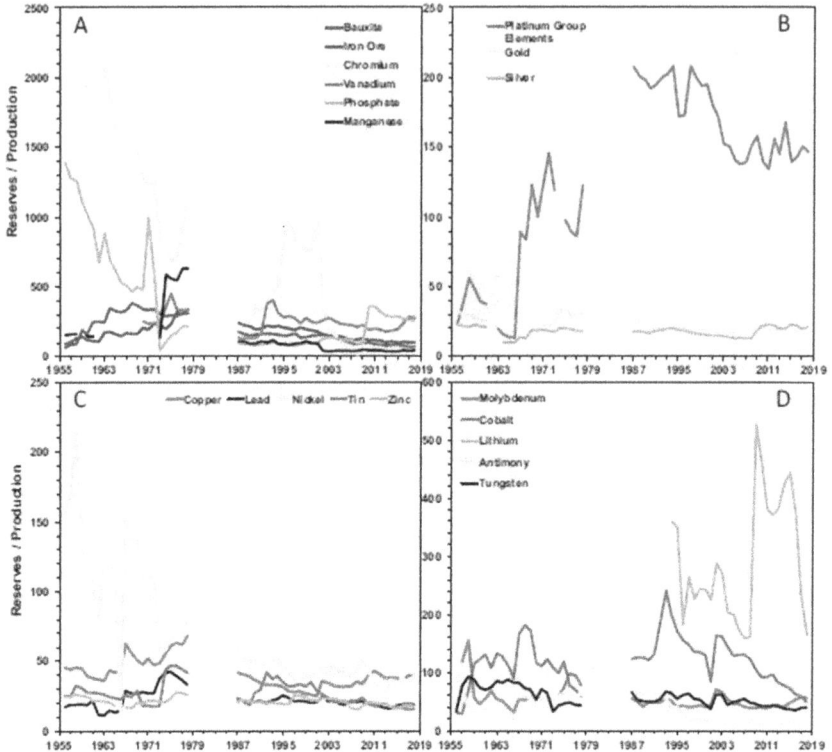

FIGURE 2.3 Annual metal production and reserve data for 1956–2018. Data are shown as reserve/production ratio for (A) selected bulk and ferrous minerals, (B) precious, (C) base and (D) minor metals. Note the data gap (1979–1986) when only "reserves base" estimates were published, which are effectively the same as resources.

Source: Jowitt, S.M., Mudd, G.M. & Thompson, J.F.H. Future availability of non-renewable metal resources and the influence of environmental, social, and governance conflicts on metal production. Commun Earth Environ 1, 13 (2020). https://doi.org/10.1038/s43247-020-0011-0; Accessed on August 14, 2021

It is evident from Figure 2.3A that ores of iron, aluminum and manganese are having a decreasing trend of R/P ratio and their metallurgical exploration may not be possible after 100 years or so. After that, primary production would be obsolete and these metals would then be available only through secondary production routes. Similarly, Figure 2.3C shows a similar fate for ore of copper in about 50 years. The variation of price of copper seems not to be dependent on either production or reserve or the ratio of R/P. The R/P ratio is increasing from 25 years in 2002 to 45 years in 2019, i.e. much better position than that of iron and aluminum.

The United States Geological Survey (USGS) further pointed out that the reserve of copper has been steady increasing since 1930 (Figure 2.5) and there is no risk of copper becoming not available in the near future; rather, it is expected that more

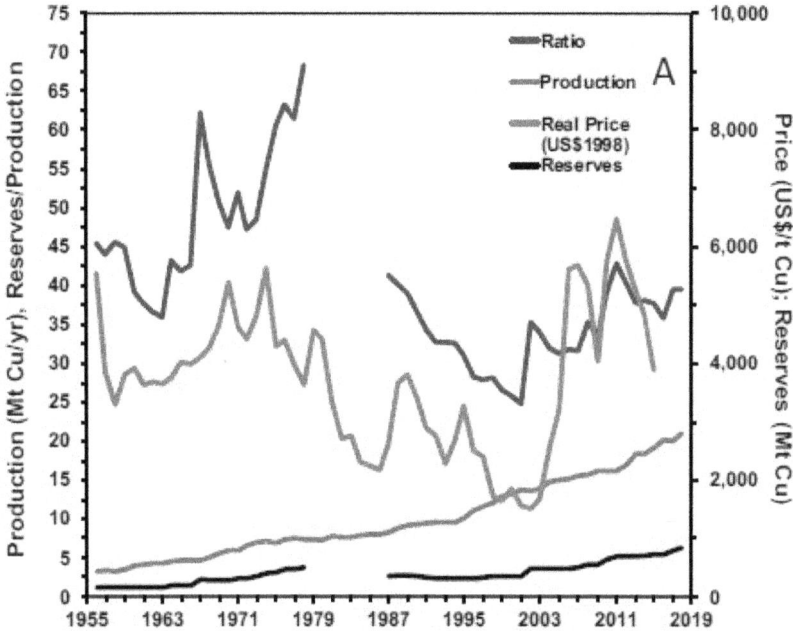

FIGURE 2.4 Copper production, reserve, reserve/production ratio and price data.

Source: Jowitt, S.M., Mudd, G.M. & Thompson, J.F.H. Future availability of non-renewable metal resources and the influence of environmental, social, and governance conflicts on metal production. *Commun Earth Environ* **1,** 13 (2020). https://doi.org/10.1038/s43247-020-0011-0; Accessed on August 14, 2021

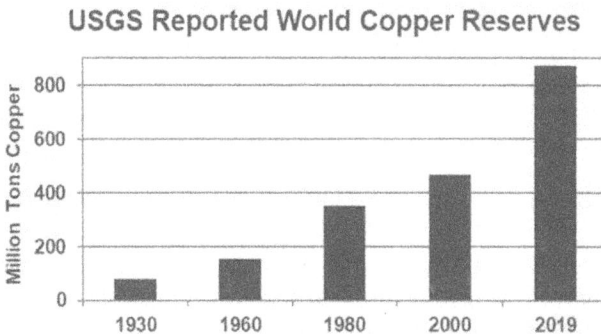

FIGURE 2.5 Trend of world copper reserve from 1930–2019.

Source: The World Copper Factbook 2020, International Copper Study Group, Lisbon, Portugal, 2020; www.icsg.org/index.php/component/jdownloads/finish/170/3046 Accessed on August 14, 2021

(undiscovered resources not including deep sea nodules and land-based and
submarine massive sulfides - contained copper)

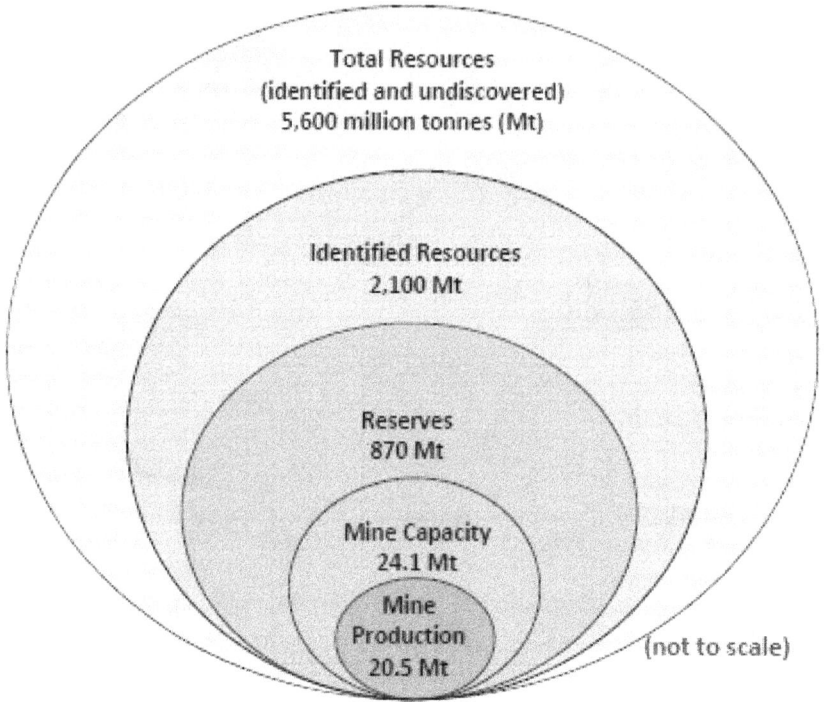

Total Resources
(identified and undiscovered)
5,600 million tonnes (Mt)

Identified Resources
2,100 Mt

Reserves
870 Mt

Mine Capacity
24.1 Mt

Mine
Production
20.5 Mt

(not to scale)

1/ Source: USGS (resources/reserves data) and ICSG (capacity/production data)

FIGURE 2.6   2019 world copper reserves and mine production.

Source: The World Copper Factbook 2020, International Copper Study Group, Lisbon, Portugal, 2020; www.icsg.org/index.php/component/jdownloads/finish/170/3046 accessed on August 14, 2021

reserve would become available as exploration continues as per available information (Figure 2.6).

Modeling tools based on the Internet of Things (IoT) and information technology (IT) are used to estimate the country-wise production of metals in different scenarios to help prepare appropriate policy based on historical and projected data as shown in Figure 2.7 (Meinert et al., 2016).

Another study revealed that demand for copper and aluminum for the global electricity system under various scenario assumptions can vary with a maximum of 2.2 MTPA (million tons per annum) and 5.7 MTPA, respectively (Figure 2.8). It was also well predicted that silver and platinum demand would be steeply rising in view of upcoming large-scale application of certain technologies (UNDP, 2013).

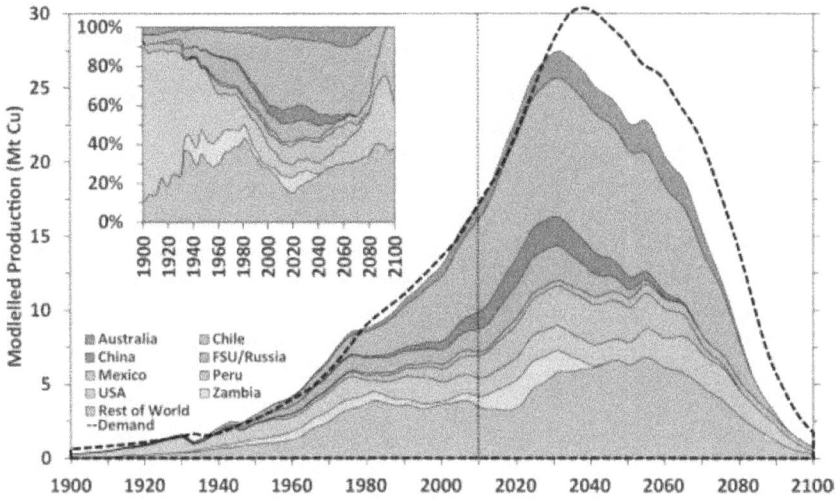

FIGURE 2.7   World copper production: historical and projected based on the modeling.

Source: Meinert et al. (2016)

Another challenge faced by metal industries is the increasing amount of GHG emissions raising environmental concerns. The global situation of GHG emission for metal extraction and refining, and mineral processing and concentration of various metals, is shown in Figure 2.9. However, with increasing share of recycled metal in metal manufacturing, improving EE&C is projecting a decreasing trend after 2040.

In "Explainer: These six metals are key to a low-carbon future" [3], it has been pointed out that in order to meet the Paris Agreement's goals of limiting global warming to "well below 2°C" and to strive for 1.5°C, fast adaptation of low-carbon technologies for the six more important metals—viz. aluminum, silver, steel, nickel, lead and zinc, along with copper, lithium, nickel, cobalt and neodymium (rare Earth elements)—will be needed.

Therefore, EE&C brings sustenance to metal industry and a sustainable metal industry will lead to innovative EE&C opportunities. The energy trilemma will converge to the common objective of a sustainable metal industry through coherence of action plan. Some of such interlinkages are presented in Table 2.2, which will lead to policy considerations for sustainable metal industry.

2.3   INTERNATIONAL AGREEMENTS AND PROTOCOLS

2.3.1   Local to Global and Global to Local

Today's metal industry at any location survives only with collaboration necessary for the flow of workers, materials and money in the entire supply chain. Its interaction

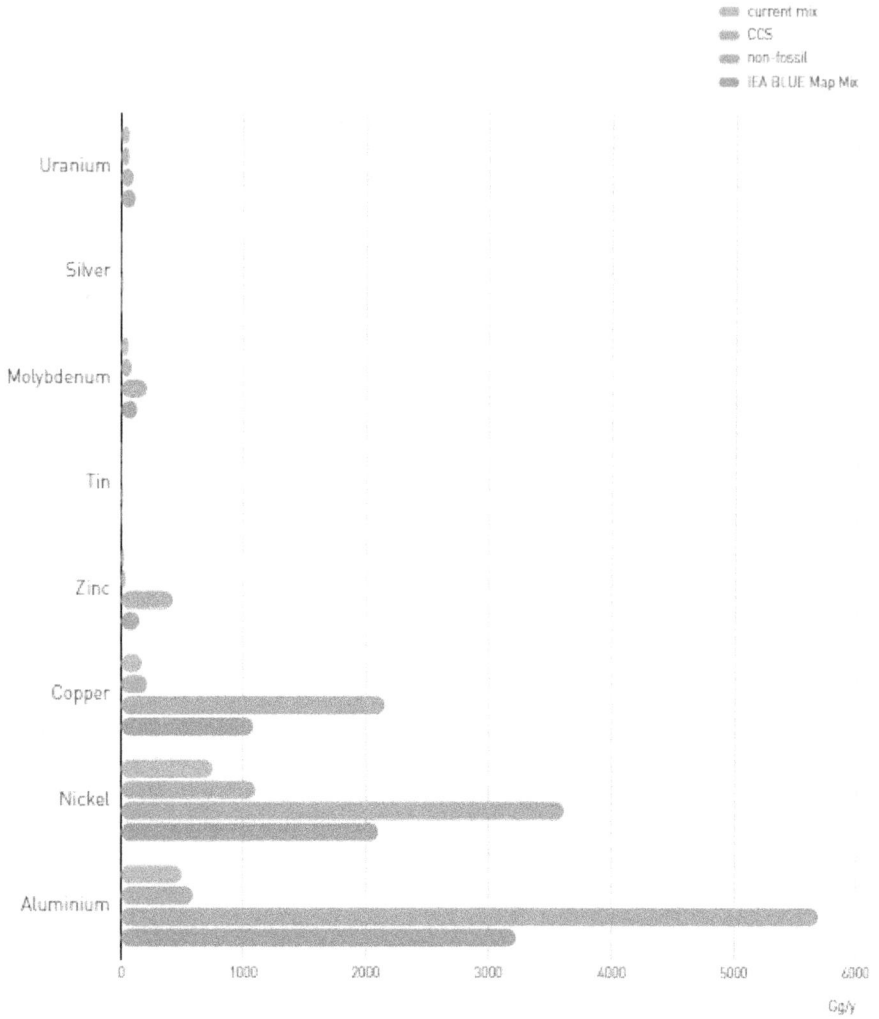

FIGURE 2.8 Demand for metals for the global electricity system under various scenario assumptions; demand for metals, such as aluminum, nickel, copper, etc., for the global electricity system under various scenario assumptions, viz. current mix, CCS, non-fossil and IEA BLUE Map mix.

Source: UNDP (2013)
Notes:
CCS: carbon capture and storage on fossil fuel–based power plants
Non-fossil: mix of solar, wind and hydropower
IEA BLUE Map: mix according to Shell Blue Map scenario, including fossil fuels as well as renewable energy sources

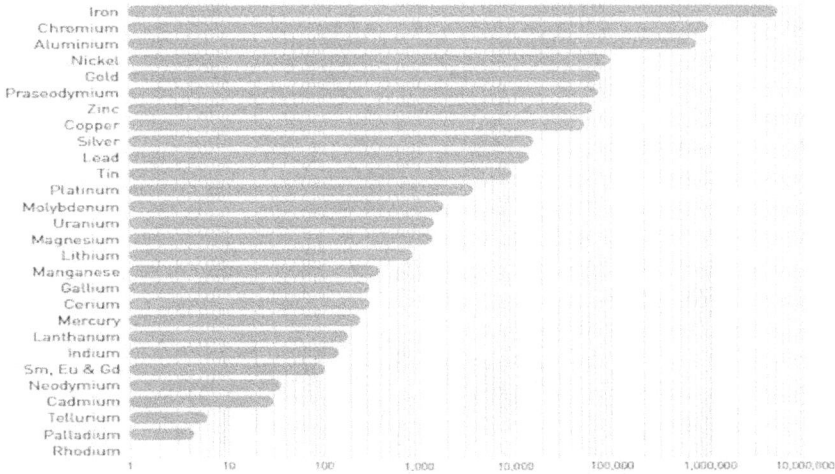

FIGURE 2.9 Contribution of global production of metals to global GHG emissions, normalized to Rh production (=1).

Source: UNEP (2013)

TABLE 2.2
Coherence of Energy Efficiency and Conservation with Sustainable Metal Industry Policy

SR No	Energy Efficiency and Conservation	Sustainable Metal Industry	Coherence
Case of a Steel Plant: Nippon Steel Corporation Group			
1	Policy in continuous persuasion to maintain world's top-class energy efficiency, innovation and implementing with top runner approach cost-effective technologies/techniques. Nippon Steel is actively involved in the Action Plans for Low-Carbon Society by promoting "three ecos and innovative technology development" and further CO_2 emission reduction.	Research and development of eco-products, with the aim to apply them commercially by 2030 through COURSE50, Carbon Capture and Utilization (CCU) and hydrogen reduction steel making scheme.	The Group is committed to adopt and continuously update with world-class technologies and application of eco-products; integrated approach of EE&C and Steel Policy framework is going to secure annual CO_2 emission reduction of 14.2 Mt by improvement of EE of steel production in Japan, of 107 Mt as EE steel materials are used in final products, of 215.5 Mt by technology transfer to the rest of the world, in three phases until 2030.

(Continued)

TABLE 2.2   Continued

SR No	Energy Efficiency and Conservation	Sustainable Metal Industry	Coherence
Case of a Country: India [4]			
2	Energy Conservation Act 2001 (Amended 2010) identifies the iron and steel industry as a designated consumer if annual fuel consumption is 30,000 tons of oil equivalent or more, meaning thereby compliance with the Energy Conservation Act which includes achieving the target specific energy consumption within three years, procure and use energy-efficient star labeled products, implement Perform Achieve and Trade (PAT) scheme, etc.	National Steel Policy 2017: The Government of India decided to foster and support an environment conducive to clean and green environment through stringent norms to attain energy efficiency and application of best available technologies.	In India, the steel sector is adopting best practices of EE&C. Sixty-seven iron and steel designated consumer units (with annual energy consumption more than 30,000 TOE) were having annual energy consumption of 25.32 MTOE, had set a target to reduce in PAT Cycle-1 (three years: 2014–2015 to 2016–2017) 1.486 MTOE, which was exceeded by more than 40% by achieving annual saving of 2.1 MTOE.
Case of a Region: European Union [5]			
3	Hydrogen Strategy, Renovation Wave, Methane Strategy, and Off-Shore Renewable Energy Strategy designed in a manner to fulfill the goals of each other, including that of reducing GHG emissions by at least 55% of the 1990 level being updated/reviewed from time to time to meet the long-term goal of becoming carbon-neutral by 2050 as also mentioned in "Fit for 55 Package" and as already agreed upon by both the houses of the European Parliament and the Council.	Zero-carbon steel-making initiatives and New Circular Economy Action Plan for Industries are two breakthrough happening in EU economy. The industry—as well as all other stakeholders, including the public—are taken to confidence through open house debate and decided to also have an intermediate target for 2040 and even to have plan to become carbon negative after 2050.	European Steel Association (EUROFER) and European Union are working hand in hand for a consistent regulatory framework whereby relevant policies are being aligned to playing complimentary roles, meeting needs to meet the goals of committed GHG mitigation, energy targets, Emissions Trading System (ETS) or the Industrial Emissions Directive (IED), as well as cost efficiency for international competitiveness. All these are leading to a successful "Fit for 55 Package."

with the environment, economy and energy within the social framework is interdependent and crossed local, regional and national borders as discussed in Section 2.2 on energy trilemma issues. World and regional bodies, such as the UNFCCC, the European Union, etc., have worked out relevant "Race to Zero" and "Steel Action

Plan" guidelines, respectively, to identify and resolve problems or overcome challenges [5]. Analysis of such international agreements and protocols illustrates the relevance and reference of regulatory framework required for the desired policy initiative. The global action plans are thus reflections of resolutions made based on cumulative local plans. If the local action plan takes care of the global counterpart, then it would be more consistent and inclusive of necessary goals for the public good.

2.3.2 Objectives and Purposes of Selected Agreements and Protocols

Conference of the Parties (COP): With an objective to "stabilize greenhouse gas concentrations in the atmosphere at a level that would prevent dangerous anthropogenic interference with the climate system," COP to the UNFCCC, at the invitation of the Government of Germany, held its first session in 1995 in Berlin, Germany.

Kyoto Protocol: This international agreement was reached in 1997 at COP III of UNFCCC and was signed by 192 countries asking for long-term definite action to mitigate anthropogenic impact on climate change. It came in to force in 2005, introducing a mechanism for carbon trading so that the developed nations can achieve respective given targets to reduce the greenhouse gas emission either within their country and/or in developing countries, ultimately globally [10].

Steel Action Plan: In the communication dated as early as 2013, the European Commission to the parliament, the Council, the European Economic and Social Committee and the Committee of Regions elaborated in the "Action Plan for a Competitive and Sustainable Steel Industry in Europe" gives a detailed outline of the industry for contribution to the environment and economy of the society through efficient and sustainable use of energy sources, as well as restructuring the legal framework and social dimension to meet the requirements of suitable manpower. The Steel Action Plan lays the strategies in the medium term and long term (2030/2050) to achieve its goals which is set in tune with energy, environment and social framework for related plants and their goals at regional and global level. Thus, the document [6] and the subsequent actions are exemplary demonstration of present visionary role the European steel industry is playing in overcoming all odds to attain carbon neutrality by 2050 and yet remain competitive internationally.

Sustainable Development Goals (SDGs): The SDGs, comprising of 17 goals, also known as the Global Goals, in continuation with Millennium Development Goals (MDGs for the years 2000–2015), were adopted by the United Nations in 2015 as a universal call to action to end poverty, protect the planet and ensure that by 2030, all people enjoy peace and prosperity. SDGs 6–9 directly and all other SDGs indirectly are related to EE&C and sustainability of metal industries in particular, as they play major or significant role toward their achievement.

Paris Climate Agreement: Signed by 196 parties at COP 21 of UNFCCC held in 2015 in Paris, this legally binding international treaty came in to force in 2016 and calls for climate action as a last best chance for anthropogenic course of corrective action on climate change so that the rise in global temperature with respect to pre-industrialization period is limited to 2°C, preferably below 1.5°C. The agreement works on a five-year cycle within increasingly stringent action on climate change by the countries. Every country has submitted its nationally determined contributions (NDCs) toward this end incorporating long-term low greenhouse gas emission development strategies (LT-LEDS), but the majority have failed to submit their reviews of more stringent carbon action plan within the deadline, i.e. December 2020, and were expected to review and address the issues at COP 26 held in end of 2021 at Glasgow. COP 26 called for action plan to phase down coal, and the governments of the United Kingdom, India, Germany and Canada have announced a pledge to purchase low carbon-emission steel under the Industrial Deep Decarbonization Initiative (IDDI). This has triggered a series of actions around the globe at various levels, including EE&C in metal industries. All the major emitters have pledged to attain carbon neutrality within the mid-century.

Race to Zero: This UN-backed initiative launched in 2019 was discussed in Section 2.2.1.

2.3.3 Countries' Preparedness

Such a huge number of signatories of the global agreements and protocols does not necessarily mean that humanity would be able to achieve the long-term goals set for the middle of this century. Rather, the response so far is far from satisfactory for majority of the signatories as reflected from their short-term and medium-term action plans befitting the long-term goals.

The metal industries, being capital intensive, look for long-term policies, especially for a stable regulatory framework; however, in order to meet the emerging goals in a changing situation, short-term and medium-term policies are also effective when situation so demands. The COVID-19 pandemic situation was one such eye-opener. Marlene Arens et al. (2021) conducted a study of which the outcome is shown in Table 2.3 and highlights are placed below.

1. The world's major 26 steel-making countries were studied.
2. All these countries ratified the Paris Agreement (except the United States, which rejoined on 20 January 2021 after a brief withdrawal).
3. All these countries also have submitted intended nationally determined contributions (INDCs).
4. Almost all these countries have declared their CO_2 reduction targets for 2030.
5. Fewer than half of them have declared renewable energy targets for 2030.
6. Even fewer (only three, until recently three more joined to make this number now six) have agreed on CO_2 reduction and renewable energy targets in the longer run, say, 2050/2060.
7. This increase of participants from three to six in recent time indicates that the rate of awareness and practical action is on an ascending path.
8. Countries and states (where they have the ability to lay their policy independently) with higher coal-based steel production base have lower participation (only account for 17% of CO_2 emissions) to such long-term declaration/commitment.

TABLE 2.3

EE&C Achievements and Metals Policy, CO$_2$ Reduction and Renewable Energy Achievements, Targets and Goals up to 2050

Sr #	Country	Renewable Power Target 2030	CO$_2$ Reduction Target 2030	Renewable Power Target 2050	CO$_2$ Reduction Goals 2050
1	China	35%	Peak/60–65% (GDP, 2005)		Carbon neutrality 2060
2	European Union	32%	40% (1990)	At the country level	Climate neutrality
3	India	40% (excl. hydro >25 MW)	33–35% GDP (2005)		
4	Japan	22 . . . 24%	−25.4% (2005, incl. land use, land-use change and forestry [LULUCF])		Carbon neutrality
5	Korea, Rep. of	20%	−37% (BAU)		Carbon neutrality
6	Russian Federation	4.5% (2020; non-hydro)	−25 . . . 30% (1990)		
7	United States	At state level	(−26 . . . −28% (2005) by 2025)	At the state level	−80% (2005)
8	Brazil	23% (non-hydro)	−43% (2005, incl. LULUCF)		
9	Ukraine	20%	−40% (1990)		−50% (1990)
10	Taiwan, China	20% (2025)	−50% (BAU)		
11	Turkey	30% (2023)	−21% (BAU)		
12	Canada	At state level	−30% (2005)		−80% (2005)
13	Vietnam	10%	−8% (BAU)	100%	
14	Mexico	35% (2024)	−25% (BAU)	50%	−50% (2000)
15	Kazakhstan	30% (incl. nuclear)	−15% (1990, incl. LULUCF)	50% (incl. nuc.)	−25% (1992)
16	Australia	23% (2020)	−26 . . . −28% (2005, incl. LULUCF)		At the state level
17	South Africa	9%	614 Mt CO$_2$, eq; (no reduction)		Peak by 2035
18	Indonesia	26% (2025)	−29% (BAU)		
19	Iran	5 GW (2020)	−4% (BAU)		
20	Argentina	20% (2025)	−15% (BAU)		
21	Algeria	27%	−7% (BAU)		
22	Serbia	37% (2020)	−10% (1990)		
23	Chile	20% (2025)	−30% GDP (2007)		Carbon neutrality
24	Bosnia and Herzegovina	40% (2025), final energy	−2% (1990)		
25	New Zealand	90% (2025)	−30% (2005)		Carbon neutrality
26	Egypt	42% (2035)	No target		

Source: Arens et al. (2021)

Thus, it may well be said that the practical actions taken by nations—particularly the bigger emitters from metal industries—are far behind the commitments made, say in the Paris Climate Agreement. Urgent action is required to actually achieve the target in 2030 and mid-century. Preferably, short- and medium-term targets should be synchronized to achieve the long-term targets, and an appropriate regulatory framework needs to be in place.

Similarly, targets for renewable electricity—the counterpart of EE&C as the instruments for attaining sustainability—are also required to be achieved through coherent short- and medium-term targets, and suitable policy in place for the metal industries; otherwise, even the most optimistic and noteworthy targets by major metal producing countries/unions, including the four mentioned in what follows, may remain unfulfilled and the purpose of sustainability of metal industry will remain a distant cry.

1. India: 40% of renewable electricity excluding large-scale hydropower (>20 MW) i.e. 523 GW by 2030
2. Egypt: 42% renewable by 2035
3. China: 35% renewable by 2030
4. EU: 32% renewable by 2030

However, a few countries, such as Japan, have been showing both short-term and long-term actions on renewables, CO_2 abatement and EE&C.

2.4   NATIONAL POLICY DRIVERS

2.4.1   THINK GLOBALLY AND ACT LOCALLY

"Think globally and act locally" is the watch phrase of national policy initiatives for EE&C in metal industries as human civilization is completely interdependent through trade and economies, and collaboration will continue to be the buzzword to overcome challenges of global warming, poverty and inequality. In the following Boxes 2.1–2.5, excerpts of national policy issues of five countries/unions—viz. India, European Union, China, Japan and Ghana—with prominent roles in metal industries are presented.

BOX 2.1   EXCERPTS ON NATIONAL STEEL POLICY OF INDIA

National Steel Policy enshrines the long-term vision of the government to give impetus to the steel sector. The policy envisions creation of a technologically advanced and globally competitive steel industry that promotes self-sufficiency in steel production, as well as economic growth. Steel being a de-regulated sector, government acts as a facilitator by creating enabling environment for development of steel sector. Various initiatives taken by the government are as follow.

1. Domestically Manufactured Iron and Steel Products (DMI&SP) Policy with an objective to encourage production and consumption of domestically produced steel.
2. Steel Scrap Policy to enhance the availability of domestically generated scrap.
3. Issuance of steel quality control orders to prevent manufacturing and import of non-standardized steel.
4. Steel Import Monitoring System (SIMS) for advanced registration of steel imports.
5. Engagement with various stakeholders, including industry associations and leaders of the domestic steel industry, to identify their issues required to be addressed by the concerned ministries/departments of the central and state governments.
6. Engagement with relevant stakeholders including those from the ministries/departments of railways, defense, petroleum and natural gas, housing, civil aviation, road transport and highways, agriculture and rural development sectors to enhance the overall demand for steel in the country.
7. Inclusion of "specialty steel" under the Production Linked Incentive (PLI) scheme recently announced by the government.
8. Various schemes as notified from time to time to refund or exempt taxes and duties levied on inputs used in export production like Duty Drawback scheme and Advanced Authorization scheme, etc., to improve the cost competitiveness of exported items.

Source: www.pib.gov.in/PressReleasePage.aspx?PRID=1704809
accessed on August 14, 2021

BOX 2.2   EXCERPTS ON THE EUROPEAN STEEL INDUSTRY

The European steel industry is a strategic sector at the heart of the EU economy—responsible for 330,000 direct jobs and 2.6 million indirectly or induced jobs overall. The 160 million tons of finished steel the industry produces every year make modern life possible, with the metal used in automotive production, construction and in the creation of household and medical goods and appliances.

The importance and size of the European steel industry to the EU economy means that ensuring the regulatory and legislative framework is as appropriate is the essential work of policy makers. This includes all facets of laws which affect the way the sector operates, including climate, competition, environment and trade policies. Only through getting all of these policies right can the sector's competitiveness be enhanced.

Taken together, the way these policy approaches interact is where the EU's industrial policy comes in: the attempt to bring order to a large rulebook and to ensure that the right support and framework exists for the European steel sector to be competitive in a world of cut-throat competition. EUROFER supports the EU's work to build an industrial policy fit for the 21st century.

The European Commission updated its industrial strategy in May 2021 to ensure that industrial ambition takes account of the new circumstances following the COVID-19 crisis, while ensuring European industry can lead the way in transitioning to a green, digital and resilient economy.

This strategy lays the foundations for an industrial policy that will support the twin transitions, make EU industry more competitive globally and enhance Europe's strategic autonomy.

Source: www.eurofer.eu/issues/industrial-policy-and-the-european-steel-market/ accessed on August 14, 2021

BOX 2.3   EXCERPTS ON NATIONAL POLICY OF CHINA

To turn the COVID-19 crisis into an opportunity and to guide economic recovery, the government has announced a new infrastructure initiative involving a shift from highly polluting, export-led manufacturing to an advanced, high-tech, service-driven economy. In September, President Xi surprised the global community by committing to carbon neutrality by 2060 and aspiring to double China's economy by 2035.

The question is how to meet these goals in addition to existing ones, including achieving peak emissions by 2030. Greater productivity has been identified as an immediate and effective solution. As achieving greater output from the same inputs essentially equates to higher energy efficiency, efficiency improvements are essential to ensure that China's energy sector can support the country's transition to a modern green economy.

In May 2020, the NDRC released "Implementation Opinions on Building a Better Development Environment to Support the Healthy Development of Private Enterprises in the Energy-Saving and Environmental Protection Sector," presenting incentives such as reduced corporate income tax for energy conservation projects, as well as energy management contracts. While this government support fortifies the market, China could attract more market-savvy commercial participants to boost energy efficiency investment and innovation.

Source: www.iea.org/commentaries/
greater-energy-efficiency-could-double-china-s-economy-sustainably

BOX 2.4   EXCERPTS ON NATIONAL POLICY OF JAPAN

1. The Japanese steel industry supports Japan's ambitious policy of achieving carbon neutrality by 2050 and it will aggressively take on the challenge to realize zero-carbon steel with the aim of contributing to the Japanese government policy. The challenge includes (1) contribution through technologies and products and (2) initiatives to reduce CO_2 emissions in steel production process (i.e., zero-carbon steel).

2. Realization of zero-carbon steel is an extremely difficult challenge and that is unlikely to be realized in a straight line. Therefore, the Japanese steel industry will explore multiple pathways to the challenge by employing every possible means, including actively ongoing efforts for the drastic reduction of CO_2 emissions from blast furnace through COURSE 50 and ferro coke technologies plus CCUS (carbon capture, utilization, and storage), development of super innovative technologies such as hydrogen-based iron making, expanded use of scrap, recovery of low- to medium-temperature waste heat, and use of biomass.

3. Challenges to develop super innovative technologies: to realize decarbonization in iron making process and zero-carbon steel, it is necessary to endeavor to develop advanced technologies, such as CCUS, under a blast furnace (reduction with carbon) method with an improved reduction ratio with hydrogen. In addition, we need to spend additional huge costs to neutralize unavoidable remaining CO_2 emissions or implement the hydrogen-based iron making, which does not generate CO_2. There is no other solution.

4. The following external conditions are required for the realization of zero-carbon steel: low-cost and stable supply of large quantities of carbon-free hydrogen and carbon-free electricity, research and development of economically rational CCUS and its implementation in society

5. Fostering public understanding that the realization of zero-carbon steel requires a large amount of costs, including research and development, capital investment and operational costs, as well as establishing a society as a whole to bear these costs—ensuring equal footing where Japan's industries are not disadvantaged in international competition—including an urgent resolution of electricity prices, which remain high by international standards; introducing additional carbon pricing measures, such as carbon taxes and emissions trading schemes, which take away resources for technological development and capital investment, will hinder technological innovation and result in preventing the realization of zero-carbon steel.

Source: www.jisf.or.jp/en/activity/climate/documents/CN2050_
eng_201210215.pdf

BOX 2.5   EXCERPTS ON NATIONAL POLICY OF GHANA

There are a number of technical options to improve the energy efficiency and the fuel demand of gold mining operations. These include improving the efficiency of compressors used for drilling, avoiding leaks in pipelines for compressed air, improving the efficiency of motors in the mills, managing the routing of trucks, replacing trucks with more fuel efficient models, improving the efficiency of the cooling and of the ventilation systems in underground operations, and improving the efficiency of lighting in underground operations. Many of these options are not only technically possible, but also economically feasible with payback times of the required initial investment of less than five years, depending on the prices for electricity or fuel. However, even economically attractive measures may not be undertaken automatically due to lack of knowledge about the technical options and a lack of awareness of the potential benefits.

In Ghana, electricity prices for large industrial users such as the gold mining industry have historically been very low, as they were subsidized by the government. Such low prices posed little incentive to implement energy-saving measures. However, the situation has been changing with constant rises in electricity tariffs.

Increased efficiency of mining operations is therefore in the interest of the government of Ghana, especially as long as the government subsidizes fossil fuel imports and electricity tariffs. The investment decisions for energy efficiency measures in mining operations are taken by private companies. However, the government may play a supporting role by setting the right regulatory conditions. This could, for example, imply incentivizing investments into energy-saving measures, while at the same time punishing inefficient operations, e.g. by discontinuing electricity subsidies, or requiring the industry to continuously improve the energy efficiency of gold production.

Source: POLICY BRIEF Low-carbon options in the gold mining industry in Ghana, https://publicaties.ecn.nl/PdfFetch.aspx?nr=ECN-O – 11-023

2.4.2   SOME POINTS TO PONDER WHILE CONSIDERING A REGULATORY FRAMEWORK

Thus, at national level, there are several points to ponder in the decision-making process of policy for EE&C and sustainable metal industries.

2.4.2.1   Subsidy vs. Directives

Many of the EE&C technologies implementation requires huge investment which may be beyond the budget of the industry but upon implementation would fetch three benefits: first, monetary savings within the factory, as well as while the final product is used; second, the associated reduction in emissions; and third, the scope of replication of the best practices so demonstrated.

Therefore, considering the huge benefit to the society in the long run, EE&C policy and metals policy often uphold the provisions for subsidy to such projects. Thus, subsidy allocates public money to a purpose which would return in greater amount in terms of public good, through market push.

Directives, on the other hand, may work in a mechanism of market push or market pull. For example, consider the case of Renewable Energy Directive (RED) in the European Union, which needs to be supported through incentives, i.e. market push, rather than the new obligations i.e., market pull (EUROFER [2021, July]).

2.4.2.2 Protection vs. Competition

The international trade, depending on the situation, applies protection to the local market at a time or for a certain commodity in one hand, and at other times or for other commodity tries to provide a policy framework to provide a level plane or equal opportunity to promote fair competition in the market. Healthy competition strengthens the market, promotes efficiency and reduces waste and non-productivity. However, until the market is matured or when the market is in distress, the government follows a policy to protect that particular market. For example, Steel Policy of the government of India calls for promotion of domestic supply of recycled iron and EU calls for right support and policy framework to protect the domestic market in the face of cut-throat competition in the international steel market, as evident from the relevant declarations put in Boxes 2.1–2.5. The government faces pressure from world bodies such as the WTO to maintain the economy as open and competitive in one hand, and also pressure from local business associations for protection of markets, on the other. For example, at Brussels, on 30 June 2020, the European Steel Association (EUROFER) and IndustriALL European Trade Union have strongly expressed their disagreement with the decision of EU to maintain tariff-free import quotas for steel in contrast to their competitor countries' directives (EUROFER [2021, June]) [7].

2.4.2.3 Negotiations vs. Restrictions

Similar to protection vs. competition, negotiations and restrictions are also useful tools used in regulatory frameworks to control the market forces for its revival and growth. Negotiations at national and international levels can reduce misunderstandings and open up channels of win–win situations. On the other hand, until such collaborations attain maturity, restrictions are applied in demanding situations to protect the interest against the threat.

2.4.2.4 Green Primary Metal Production vs. Recycling

Recycling of metals have multiple advantages, as discussed in this chapter. Until 100% recovery is attained and while the demand is greater than supply of recycled metals, green primary metal production will continue, subject to availability of ores. Policy needs to take care of optimum utilization of installed capacity and other factors of production.

2.4.2.5   Total vs. Per Capita

The United States and many developed countries call for treating India and China as countries with very high rates of emission. This is because in developing economies like China, India, Brazil, Argentina, etc., rapid industrialization—particularly carbon-intensive production—has been taking place along with often increasing population. As shown in Figure 2.10, more than 90% of the world's iron and steel has been produced in six countries/regions—China, the European Union, India, Japan, South Korea and the Russian Federation in 2018. Further, the world's top ten steel producing countries in 2020, listed in Table 2.1, also are among the highest emitters of CO_2. China tops the list in both steel production and CO_2 emissions, the United States is second to China in CO_2 emission but fifth in steel production, and India is third in emissions and second in steel production among countries of the world. This increasing rate of emission in developing economies is a matter of great concern to the global planners of EE&C, and sustainability of metal industries, too.

The idea of global common and rights to equality and development of individuals is the basis of human civilization. It is interesting to note that the ranks among those top ten producers of steel become topsy-turvy when per capita values are calculated—for example, India ranked second in the world after China in terms of steel production, but, in terms of per capita production, attains the last rank among these ten countries. China slips from first to third and South Korea moves up from sixth to first position. Also, the rank among these ten top producers of steel in terms of per capital production and per capita CO_2 emission are not much different—for example, India maintains the last rank in both, followed by Brazil at ninth in both. South Korea, which topped in per capita steel production, is second to the United States in per capita CO_2 emission. The matter of concern is the direct relationship of per capita CO_2 emission with that of human development index (HDI) of these ten countries;

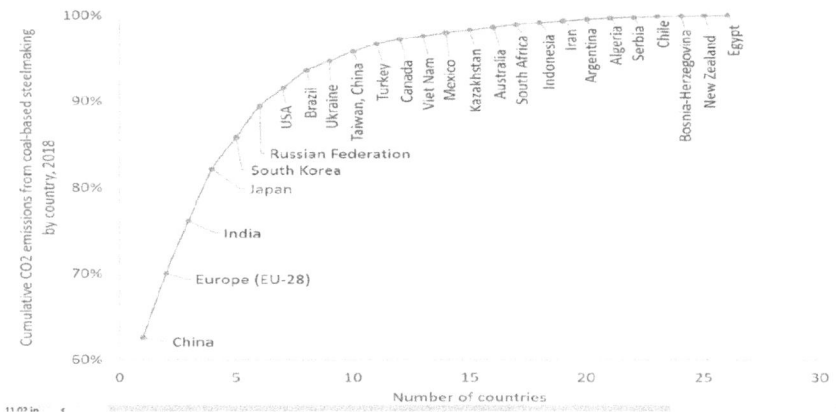

FIGURE 2.10   Cumulative CO_2 emissions from coal-based steel making by country, 2018. Assumptions: blast furnace steel making 2 t CO_2/t steel; coal-based direct reduction 3 t CO_2/t steel (mainly in India).

Source: Arens et al. (2021)

the ranks are very close—for example, China, 7 and 9; India, 10 (last) in both; Japan, 4 and 3; Russia, 3 and 5; and USA, 1 and 2, respectively, for per capita CO_2 emission and that of HDI. In fact, the difference in ranks is less than three for all these ten countries except that of Germany ranked first in HDI but fifth in terms of per capita CO_2 emission. This global concern has culminated in tremendous research, planning and policy to delink carbon from economic development, evolving the concept of low-carbon economic development, hydrogen economy, etc., and technologies like HYBRIT, green hydrogen driven fuel cells, green production, and so on.

2.5 EE&C POLICY CONSIDERATIONS: TIPS ON THE TOP

2.5.1 EE&C POLICY NEEDS TO BE TECHNOLOGY SAVVY

Innovations can change the entire development path. Some of the innovations of metal production are discussed in Chapter 4, such as HIsarna Technology producing iron from its ore without use of coke and with almost 100% CO_2 concentration in the off-gas which can be ideally used as raw material or carbon sequestration; HYBRIT Technology producing steel with use of 100% renewable energy; and many other technologically viable processes which may soon be available is market as commercial viable options.

On the other hand, the applications are also going through path-breaking innovations [8] which may change the scenario and subsequently the policy, some of which are mentioned in what follows:

1. Inflatable steels are the new technology-savvy product, being much lighter, logistics friendly and yet cheaper. For example, when it would be commercially available, the cost of each 2-meter long wind turbine blade—which are nowadays made of carbon fiber—available at over USD 700 will be made of inflatable steel and available at less than USD 30.
2. Metals in packaging will be durable and contribute to the circular economy. If the manufacturers are the owners of the containers and users are only borrowers, then metals can replace plastics in many cases—and thus also reduce the environmental damage caused by plastics. Addition of value and facilities will also make such containers more attractive and effective.
3. Light steel framing for innovative building technology can make it greener, safer and faster to construct, particularly effective in need of emergency, for example, rapid construction of a healthcare facility/COVID-19 hospital during the pandemic. Such technology also may pave the way for green building to be cheaper.
4. Metallic hyperloops may revolutionize e-commerce, which may be able to restrict the emissions of last-mile delivery systems in a business-as-usual scenario, and is going to increase by 30% in world's 100 largest cities by 2030, according to a study by the World Economic Forum (2020).
5. As electricity costs from utility-scale solar PV fell 13% year-on-year, reaching approximately seven cents (USD 0.068) per kWh in 2019, and from onshore and offshore wind both fell about 9% year-on-year, reaching approximately five cents (USD 0.053) and 12 cents (USD 0.115) per kWh, respectively, in the last 50 years [11], many technologies,kk—such as green steel production, carbon capture from air and storage underground

(in basalt rock where over the course of a few years to become mineralized, or utilize it as raw materials in beverage or fertilizer industry or in a greenhouse, along with advent of metallurgy)—would become commercialized and change the scenario. Metals, with appropriate policy initiatives, will play vital roles to make this project a success.

2.5.2 EE&C POLICY NEEDS TO BE PRAGMATIC

Let us consider a car having a useful life of 15 years, used 960 kg of steel produced from scrap @2.75 kWh/kg steel consuming a total of 2640 units of electricity, i.e. @ (2640/15 =) 176 kWh/year. Thus, a land area of 15.8 m^2 will be required to be cultivated to produce the required biomass. Therefore, if policy of a country tries to utilize the biomass energy sources to supply the projected demand of steel sector, it has to look in to the availability of land for the purpose in the vicinity of the proposed industry sector. Different policy scenarios for projection of cardinal parameters, such as emission from steel production, have been reported by International Energy Agency as shown in Box 2.6. It may be noted that in a pragmatic approach, three scenarios have been taken in to consideration—business as usual (BAU), i.e. the present case as in 2019; stated policies scenario (SPS); and sustainable development scenario (SDS). The emission intensity (direct, including indirect) of steel production has been projected for the year 2050 in the three scenarios as (1.4, 2.0), (1.1, 1.5) and (0.6, 0.8) CO_2/t, respectively.

BOX 2.6   CONSIDERATION OF VARIOUS SCENARIOS FOR ECONOMIC DEVELOPMENT OF THE WORLD TO PROJECT CARDINAL EE&C PARAMETERS OF METAL INDUSTRIES, VIZ. CO_2 EMISSIONS FROM THE STEEL INDUSTRY

To the extent that much of the existing capital stock will still be in operation decades into the future, the associated CO_2 emissions are often considered to be "locked in." However, these emissions are by no means destined to take place, and there are several strategies and technologies that can be deployed to varying extents to help "unlock" emissions from existing infrastructure, as follow.

- Early retirement or interim under-utilization of assets, either because of a change in policy or market conditions that makes them uneconomic or because of laws and regulations that force early closure or partial operation.
- Refurbishment and retrofitting, such as enhanced process integration to boost energy efficiency, or the application of emission reduction technologies such as replacing natural gas by hydrogen or applying CCUS.
- A change in material inputs, for example a higher share of scrap use in various process units, or higher-quality iron ore, although both of these options are limited by availability.

- Fuel switching and incremental blending, sometimes combined with some degree of retrofit, to allow assets to use less carbon-intensive or recovered fuels.

In addition to the nuances at the sub-sector level, the scope for unlocking emissions varies greatly across regions, according to the age of the different types of infrastructure. In the regions where industrial capacity is generally older, there is much more potential for early retirement, as the economic losses involved would be significantly lower. In the countries with younger assets, greater emphasis is likely to be placed on retrofitting with more energy-efficient and less carbon-intensive technologies, where it is economic to do so.

Beyond applying the mitigation strategies mentioned previously, existing production facilities can be used to bridge the gap to breakthrough technologies. This is especially important for the sustainable transition of the steel sector, where readily available alternatives for dramatic reductions in emission intensity are not commercially available today.

Strategically timed investment to partially renew existing infrastructure—or a decision to forgo investment—can form an important strategy to avoid a new investment cycle occurring just at the wrong time. By eliminating the seemingly small period of 15 years between the 25-year investment point and the typical 40-year plant lifetime, operators can have a big impact on future emissions from the iron and steel sector: around a 30 Gt CO_2 reduction in cumulative emissions from these assets, or approximately 50%.

. . . .

Steel production in advanced economies remains relatively stable through 2050, while declining markedly in China, the single largest driver of past global growth. India drives world production growth to 2050 as output rises by threefold to fourfold by 2050 in both scenarios.

Following China, the leading steel producers in 2019 included the European Union (9% of global production), India (6%), Japan (5%), the United States (5%), Korea (4%) and the Russian Federation ("Russia") (4%). Considerable growth in steel production in India is expected in the coming years, driven by economic development and the government's stated intention to build up the nation's steel industry. This growth would be in line with its ambitions under the "Make in India" initiative to transform the nation into a global manufacturing hub (Chapter 2). India's production increases nearly fourfold by 2050 in the SPS, and threefold in the SDS. This brings India's production in both scenarios to 17% of global production, significantly reducing the dominance of China. Thus, India's pathway is a critical component of any sustainable transition in the steel sector.

. . . .

In the SPS, the BF-BOF route remains the dominant pathway for producing steel, with around 250 MTOE of off-gases being generated in 2050. About 60% of these off-gases are used to fulfill onsite heat requirements (their emissions are considered direct emissions) and the remainder is used to produce power for the steel sector (their emissions are considered indirect emissions). Global direct emissions are 2.7 Gt CO_2 in 2050, but when adding indirect emissions, the figure rises by over 40% to 3.9 Gt CO_2. This means that the steel industry's contribution to global energy sector emissions is projected to be around 7% on a direct emissions basis in 2050, and 10% when including indirect emissions—very similar shares as today.

In the SDS, the technology portfolio undergoes a radical shift, with widespread deployment of production pathways that either manage the carbon contained in these gases once it is generated (e.g. by deploying CCUS) or avoid the generation of off-gases in the first place (e.g. switching to hydrogen-based production). By 2050, the total generation of off-gases is 130 MTOE, about 50% lower than in the SPS, and 40% lower than in 2019. This implies greater use of onsite generation using other fuels (e.g. natural gas or bioenergy), increased supply through dedicated renewable power or greater reliance on imported grid electricity. In the SDS, the electricity supply (excluding that supplied by off-gases) de-carbonizes by over 95%, from 540 gCO_2/kWh on average in 2019 to 18 gCO_2/kWh in 2050.

The change in emission intensity in each scenario is marked. From a direct emission intensity of 1.4 t CO_2/t in 2019 (2.0 t CO_2/t including indirect emissions), it declines to 1.1 t CO_2/t in the SPS by 2050 (1.5 t CO_2/t including indirect emissions). This reduction is due to a higher proportion of scrap as a share of total metallic inputs. In the SDS, the decline is much steeper, reaching 0.6 t CO_2/t by 2050 (0.8 t CO_2/t including indirect emissions).

. . . .

Emission reductions are measured relative to the SPS; as such, the proportion of improvements relative to today that occurs in both scenarios is not represented (e.g. a significant share of increases in scrap-based production). Material efficiency here refers specifically to demand reduction. Electrification here includes only direct electrification, primarily via conventional technologies, including shifts toward secondary production in EAFs and electrification of ancillary process equipment like preheaters and boilers. Hydrogen here refers specifically to electrolytic hydrogen, while so-called blue hydrogen (via natural gas–based DRI with CCUS) is included under CCUS. Other fuel shifts include primarily coal to natural gas switching.

Technology performance improvements and material efficiency deliver 90% of annual emission reductions in 2030. In the longer term, innovative technologies such as carbon capture-equipped and hydrogen-based production are required for further emission reductions.

Source: IEA (2020), Sustainable Recovery, IEA, Paris www.iea.org/reports/
sustainable-recovery (Selected portion)
Notes: SPS = stated policies scenario; SDS = sustainable development
scenario

2.5.3   EE&C Policy Needs to be Coherent

Every government nowadays has hundreds of policies, many of which might have transformed in law. Any new policy, before finalization, must be reviewed exhaustively to find if any section or clause in it contradicts any other within the said policy or any other existing policies. Efficiency of the good governance is much improved with coherence of policies, which support each other's goals directly or indirectly. For example, energy efficiency policy and renewable energy policy should corroborate each other's goals, even also with that of finance policy, so that the projects and activities get boosted up with the positive outlook from the relevant sections of the policies for its success.

2.5.4   EE&C Policy Needs to be Short-/Medium-/Long-Term

Depending on situation, policy needs to be short term or medium term or long term, which are generally relative concepts, typically of the span of up to one year, 1–15 years or more than 15 years, respectively. Policies to tackle the immediate effect of the COVID-19 pandemic, for example, need to be short term—but to develop a restoration plan could be a medium-term policy and preventive mechanism for any such pandemic in future could be a long-term policy.

2.5.5   EE&C Policy Needs to be Evolving Yet Binding

In many cases, it has been observed that certain policy comes out to be extremely useful and/or popular. Such a policy can be transformed to become law and then all its clauses become binding to the authority.

2.5.6   Suggestive Flow Chart for Preparation of an Effective Policy for EE&C in Metal Industries

1. Identify objective of purpose.
2. Articulate mission and vision statement.
3. Define the problem(s) and challenge(s) to overcome.
4. Identify the stakeholders.
5. Lay down what the intended impact on the stakeholders.
6. Outline the desired outcomes.
7. Prepare the draft policy.
8. Ensure that no elements of the policy contradicts existing policies and regulatory frameworks.
9. Look into the coherency of the policy with the UN and such bodies' policies.
10. Determine the mitigation plan of possible negative impacts of the policy on stakeholders.
11. Critical review of the draft policy by the concerned legal entities and administrative departments.
12. Circulate the draft policy among the stakeholders and experts, and make public notification for seeking feedback.

13. Organize a brain-storming session to review the draft policy and the feed-back from the stakeholders and experts to modify the policy.
14. Repeat steps 8–13 until a satisfactory state is attained.
15. Final approval of concerned authorities.
16. Publish the policy in public domain and circulate it among concerned department and parties for necessary action for its implementation, includ-ing the scope of its transformation to a law.

TABLE 2.4
Policies That Are the Basis for or Affect the EE&C in Metal Industries at Various Levels

Policy	Local: Factory, Municipality; Tata Group	National: State; Country: Germany	International: Region, World; European Union
Climate Change Policy	The Tata Group has over the years, since 2007, worked on its climate aspiration and strategy to be a global leader in combating climate change. This is being achieved by integrating climate change issues with business strategy by focusing on three key aspects: mitigation, adaptation and responsible advocacy.	The goal of German energy policy is to reduce such emissions by at least 40% by 2020 and by 80–95% by 2050, relative to 1990 levels.	The Commission's proposal for the first European Climate Law aims to write into law the goal set out in the European Green Deal—for Europe's economy and society to become climate-neutral by 2050.
Energy Efficiency Policy	Most Tata companies have adopted energy-efficient systems such as lighting, variable frequency drives for motors, star-rated appliances and waste heat recovery systems.	In 2010, the German government adopted an ambitious energy infrastructure transformation program that sets a long-term strategy for German energy and climate policy.	The Commission proposes a binding EU-wide target of 30% for energy efficiency by 2030, emphasizing the EU's commitment to put energy efficiency first.
Renewable Energy Policy	In 2016, Tata Motors joined the RE100 initiative and committed to transitioning to 100% renewable electricity by 2030.	The German government initially planned to further increase the share of renewables in electricity to 50% by 2030, 65% by 2040 and 80% by 2050.	Renewable Energy Directive: EC proposal to raise target for 2030 to 40%.
Metals Policy	N/A	It is of outstanding importance to have a steel industry in Germany which is strong, internationally competitive and climate-neutral on a long-term basis.	In order to maintain competitiveness, metals companies must also be proactive in anticipating and strategically planning for the challenges that Europe will face over the next 35 years.

Policy	Local: Factory, Municipality; Tata Group	National: State; Country: Germany	International: Region, World; European Union
Source(s)	Tata Sustainability Group, www.tatasustainability.com/ Environment/ClimateChange	www.umweltbundesamt.de/ en/topics/climate-energy www.iea.org/reports/ germany-2020	https://ec.europa.eu/ clima/policies/eu- climate-action/law_en www.eceee.org/policy- areas/EE-directive/ https://ec.europa.eu/ energy/topics/ renewable-energy/ directive-targets-and- rules/renewable- energy-directive_en https://eurometaux.eu/ media/1523/full-lt- sustainability- framework-document- approved-1.pdf

Instead of placing here the cases of failures, the authors would like to point out to the statement by Mahatma Gandhi: "The Earth has enough for everyone's need but not for everyone's greed." Through proper planning and implementation, all the needs for an efficient sustainable metal industry can be met only with an appropriate policy framework.

2.6 CONCLUDING REMARKS ON EE&C POLICY

All policy frameworks have some common characteristics that lead to success, such as mission/vision/goals, targets, stakeholders' benefit, satisfaction, coherence of plan of action involving and/or addressing major stakeholders' issues, identification of barriers and their overcoming, or an inherent PDCA (plan, do, check and act) cycle to help it evolve to consider and counter all dynamic issues. However, EE&C policies are special in some ways, aptly revealed to the society through the current pandemic. EE&C policy can affect the stakeholders in the short term, medium term and long term to affect the day-to-day life to the extreme conditions that the planners can imagine but cannot ignore.

The COVID-19 pandemic has hindered the implementation of many EE&C policy initiatives now and in the recent past, and will continue to affect in foreseeable future. It is estimated that the global energy intensity is expected to improve by only 0.8% in 2020, roughly half of the weather-corrected rate for 2019 (1.6%), while global investment in efficiency is projected to fall 9% in 2020.

But such a situation also gives a unique challenge and opportunity to learn from "what happens when" some unforeseen situation is bestowed on human life—as if this pandemic situation is also a part of the PDCA cycle of which half of the cycle happened, with or without our knowledge, to our civilization which is now struggling to control the remaining half. Everything depends on our appropriate EE&C action plan to catapult to a new horizon, by employing the lessons from this disaster, which may prove to be an eye-opener to the planners of imminent future and modern times.

Policy is therefore a crucial instrument for bringing the desired benefit to the society by proper growth and sustenance of the metal industry at local, national and global levels. Its appropriate deployment in short, medium and long terms in a legal framework with commitment from governments and/or multi-government unions and top management of metal industries, as well as participation of stakeholders at all levels, will remain the fundamental condition for its success.

REFERENCES

Arens, Marlene, Max Åhman, and Valentin Vogl (2021); Which Countries Are Prepared to Green Their Coal-Based Steel Industry with Electricity? – Reviewing Climate and Energy Policy as Well as the Implementation of Renewable Electricity. *Renewable and Sustainable Energy Reviews*, Volume 143, 2021, 110938, ISSN 1364–032+C2601, https://doi.org/10.1016/j.rser.2021.110938; www.sciencedirect.com/science/article/pii/S1364032121002306

Elbaum, Bernard (2007); *How Godzilla Ate Pittsburgh: The Long Rise of the Japanese Iron and Steel Industry, 1900–1973*. Elbaum, Bernard, 2007.

EUROFER (2021, June); www.eurofer.eu/press-releases/eu-steel-safeguard-decision-could-nip-european-steel-recovery-in-the-bud/

EUROFER (2021, July); Press release A fine balance: Fit for 55 must help decarbonisation of EU steel and prevent carbon leakage effectively; www.eurofer.eu/assets/press-releases/a-fine-balance-fit-for-55-must-help-decarbonisation-of-eu-steel-and-prevent-carbon-leakage-effectively/20210715-Press-release-Fit-for-55-V5.pdf

European Commission, (a); *Facts and figures on life in the European Union*; https://european-union.europa.eu/principles-countries-history/key-facts-and-figures/life-eu_en#:~:text=The%20EU%20covers%20over%204,country%20and%20Malta%20the%20smallest. accessed on April 1, 2022.

European Commission, (b); *Challenges Faced by the EU's Steel Industry*; https://ec.europa.eu/growth/sectors/raw-materials/industries/metals/steel_en accessed on August 15, 2021

European Commission, (c); *Challenges Faced by the EU Non-Ferrous Metals Sector*; https://ec.europa.eu/growth/sectors/raw-materials/industries/metals/non-ferrous_en

European Copper Institute (2013); www.eurocopper.org, https://d396qusza40orc.cloudfront.net/metals/3_Environmental_Challenges_Metals-Full%20Report_36dpi_130923.pdf#96

Meinert, L.D. et al. (2016); N.T. Mineral Resources: Reserves, Peak Production and the Future. *Resources*, Volume 5, 14; https://doi.org/10.3390/resources5010014

Nilsson, Anna Ekman et al. (2017); *A Review of the Carbon Footprint of Cu and Zn Production from Primary and Secondary Sources*, Minerals, MDPI, 2017

S&P Global (2019); *ESG Industry Report Card: Metals and Mining* – Report by S&P Global, 2019.

UNDP (2013); Environmental risks and Challenges of Anthropogenic Metals Flows and Cycles; Report #3 of the Global Metal Flows Working Group of the International Resource Panel of UNEP; 2013.

UNDP (2020); http://hdr.undp.org/en/content/latest-human-development-index-ranking

UNDP (2021); www.un.org/en/chronicle/article/icelands-sustainable-energy-story-model-world.

World Economic Forum (2020); *The Future of the Last-Mile Ecosystem*, World Economic Forum, Geneva, Switzerland, www.weforum.org.

World Energy Council (2015); *World Energy Trilemma: Priority Actions on Climate Change and How to Balance the Trilemma*, World Energy Council, 2015; www.worldenergy.org/assets/downloads/2015-World-Energy-Trilemma-Priority-actions-on-climate-change-and-how-to-balance-the-trilemma.pdf accessed on August 15, 2021.

World Steel Association (2012); *The White Book of Steel*. World Steel Association, 2012; www.worldsteel.org/en/dam/jcr:7b406f65-3d94-4e8a-819f-c0b6e0c1624e/The+White+Book+of+Steel_web.pdf

WEBSITES

[1]   https://en.wikipedia.org/wiki/Recycling]

[2]   www.apec.org/Publications?Category=B803C671F169465F851AC741331F5ED2

[3]   www.carbonbrief.org/explainer-these-six-metals-are-key-to-a-low-carbon-future

[4]   https://steel.gov.in/sites/default/files/draft-national-steel-policy-2017.pdf

[5]   www.eurofer.eu/issues/climate-and-energy/

[6]   https://eur-lex.europa.eu/legal-content/EN/TXT/PDF/?uri=CELEX:52013DC0407&from=EN

[7]   https://ec.europa.eu/growth/sectors/raw-materials/industries/metals/non-ferrous_en

[8]   https://stories.worldsteel.org/

[9]   https://data.worldbank.org/ accessed on March 1, 2022

[10]   https://unfccc.int/

[11]   https://www.irena.org/publications/2020/Jun/Renewable-Power-Costs-in-2019 accessed on April 1, 2022

3 Energy Efficiency and Conservation Technologies

Jitendra Saxena and Binoy Krishna Choudhury

3.1 INTRODUCTION

Technology is defined as the application of scientific knowledge for practical purposes, especially in industry, for benefit of the humankind. Major metal industries including iron, steel, aluminum and copper have ever-progressive technologies to extract the metal from its ore in a cheaper, efficient and more economic process. Energy efficiency and conservation (EE&C) has been gaining increasing importance in this rapidly changing scenario. It is being utilized as a tool bridging the gap between theory and practice, target and actual, productivity and sustainability, and so on.

Let us take the case of overall end-use energy through various stages of coal burning in power plants, transmission of generated electricity to user and end-use as light by him/her (Figure 3.1). Generally, efficiency (E) is expressed as ratio of desired output to the necessary input—a non-dimensional number, both numerator and denominator are expressed in the same unit of energy, limiting its value between 0 and 1. In case of electricity being used to light our home, typical efficiency for LED lights would be about 0.25, or 25%. Whereas, that electricity is generated at, say, coal-based thermal power plant, received through transmission and distribution system with an average loss of 20%, i.e. efficiency 0.8 or 80%. If the thermal power plant is producing electricity from coal at an average efficiency of 0.35 or 35%, the overall efficiency or end-use energy efficiency of the entire process of producing light energy at home by burning coal at power plant can be calculated by multiplying the efficiency of these three systems which are connected in series (from coal to light) = $0.35 \times 0.80 \times 0.25$ = 0.07, or 7%. This value is pretty low and practically could be even lower, depending on how effectively we are using the final product—in this case, light energy—which in turn, contributes to raise the GDP of the country, depending on the productive use of the final energy. This chapter also explains the effect and relevance of energy conservation (EC) vs. energy efficiency (EE), which are distinctly separate but related terms. It has been shown in this chapter that EC may not necessary improve thermodynamic EE, but would improve the EE calculated based on economic perspective.

The highest efficiency achieved in fossil fuel–based power plants converting thermal energy of fuels in to electricity is 63% at Chubu Electric Nishi-Nagoya

DOI: 10.1201/9781003157137-3

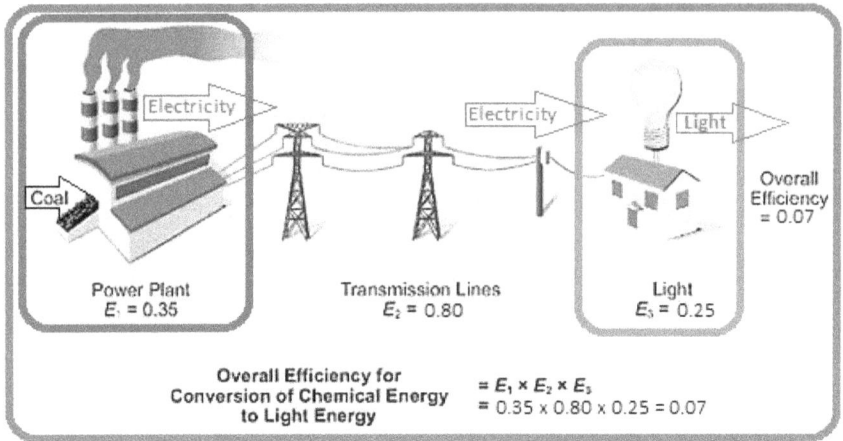

FIGURE 3.1   Concept of end-use energy efficiency.

power plant Block-1 in Japan, according to the Guinness Book of World Records [1]. Considering the common flame temperature of natural gas as 1960°C and environment at an average of 30°C, the highest possible efficiency given by Carnot as (1960 − 30)/(1960 + 273) = 86%; highest T&D efficiency achieved by Singapore at 98% (with advent of superconductor, it can rise further); and highest lighting efficiency 29%. Thus, highest overall efficiency could be 0.86 × 0.98 × 0.29 = 0.24 or 24%. Thus, it looks like we could meet all our needs at only one-third of the current energy source consumption rate, if our energy systems could operate at highest possible efficiency, which seems to be a technological challenge. Chemical, thermal, mechanical and electrical limits to efficiency are extremely relevant in energy management of metal industries for many reasons. Taking this as an example, if misuse of lighting is achieved by reducing the burning hours, this would lead to EC, but end-use EE would remain the same, or may even decrease due to part load of the power plant, if proper load management is not taken in to consideration, as will be explained in Chapters 3 and 4.

However, our experience, particularly since industry revolution, is worthy of finding many interesting features of our technological/economic/societal/environmental linkage that has a far-reaching impact on our human civilization, where role of EE&C in metal industries are becoming increasingly important.

EE&C may even lead to increasing consumption of resources due to uncontrolled expansion of economic activities, aptly stated in the Jevons paradox in 1865 and subsequently much deliberated while linking EE&C to sustainable development (Jevons, 1865).

3.2 THERMODYNAMICS AND EE

Thermodynamics is the science of energy transfer and its effects on the physical properties of substances. Thermodynamics is also said to be the science of three

Es: energy, entropy and equilibrium (Nag, 2009) that enable us to systematically study any energy system to improve its performance as expressed by several energy performance indicators, most importantly, energy efficiency. A system is defined as a quantity of matter or a region in space upon which attention is concentrated in the analysis of a problem. The system is separated from the surrounding by the system boundary, which can be fixed or moving. There may be energy transfer or mass transfer, or both, across the system boundary. Thus, thermodynamics helps us to conceptualize vast and complicated system in to simple scientific representations, such as, system + surroundings = universe, summation of all mass flow in to the system is summation of all mass outflow from the system (the law of conservation of mass), law of conservation of energy, etc., as shown in Figure 3.2. Thus, the energy balance equation for a steady-state steady-flow process, which is never happening but assumed in most cases of metal industries for energy analysis, can be written as follows.

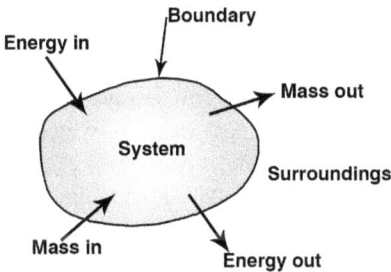

System + surroundings = universe Eq. 3 1

System in steady state (uncommon in reality but used more in practice):

Σmass in = Σmass out (=0 for closed system) Eq. 3 2

Σenergy in = Σenergy out (=0 for isolated system along with above equation) Eq. 3.3

System in unsteady state (common in reality but used less in practice):

Σmass in = Σmass out + Δmass$_{System}$ (final mass – initial mass) Eq. 3.4

Σenergy in = Σenergy out + Δenergy$_{System}$ (final energy –initial energy) Eq. 3.5

For the open system, steady-state steady-flow (SSSF) energy equation in Thermodynamics:

$$w_1\left(h_1\frac{V_1^2}{2}+Z_1g\right)+w_2\left(h_1\frac{V_2^2}{2}+Z_2g\right)=\frac{dQ}{d\tau}$$

$$=w_3\left(h_3\frac{V_3^2}{2}+Z_3g\right)+w_4\left(h_1\frac{V_4^2}{2}+Z_4g\right)=\frac{dW_x}{d\tau}$$

Eq 3.6

Where, W_1 and W_2 are mass flow rates at inlet points 1 and 2, and W_3 and W_4 are at outlet points 3 and 4, respectively. The variables h, V and Z are corresponding enthalpy per unit mass, velocity and elevation with respect to a reference line. Q and W_x are the heat and work transfer expressed in rates by differentiating with respect to time τ, d being cut with a "–" to remember that they are path function, i.e. their value depends on path being followed. The number of terms on either side may increase or decrease depending on situation of any particular system under study in metal industry. Here, h represents sum of internal energy (due to temperature) and flow work, V is for kinetic energy, Z for potential energy, Q for thermal energy transfer and W for shaft work transfer. Through experience and logic, one can develop the skill of reducing the general form of equation to a specific application neglecting the terms not so relevant/significant and incorporating which are relevant/significant.

FIGURE 3.2 Thermodynamic open system is the most common case of metal industries whereby both mass and energy interactions take place across the system boundary.

Some of the relevant thermodynamic terms and discussions are stated in what follows for further reference.

- Thermodynamic properties are certain characteristics of a systems by which its physical condition may be described, e.g. volume, temperature, pressure, etc.
- When all the properties have definite values, the system is said to exist in a definite state. Any operation in which one or more properties of a system changes is called a change of state. The succession of states passed through during a change of state is called the path of the change of state. When the path is completely specified, the change of state is called a process, e.g. isobaric process, isochoric process, isothermal process, etc.
- A thermodynamic cycle is defined as a series of state changes such that the final state is identical with the initial state, e.g. Carnot cycle, Rankine cycle (steam power plants), Otto cycle (petrol engines), diesel cycle (diesel engines), Brayton cycle (gas turbine cycle).
- Zeroth Law of Thermodynamics: If two substances are each in equilibrium with a third substance, then the two substances are in equilibrium with each other. Outcome includes thermodynamic temperature scales.
- First Law of Thermodynamics: "Energy and materials are always conserved." Outcome includes equations of energy and mass balance, which are said to be the law of conservation of mass and the law of conservation of energy. The sum total of mass (and energy) remain constant in an isolated system and in the universe, as long as the nuclear reaction, represented by $E = mC^2$, is not taking place. In a nuclear reaction, loss of mass (rest mass) times (velocity of light in free space squared) equals the amount of (thermal) energy released. Therefore, in the absence of a nuclear reaction, energy can be neither created nor destroyed, and only can change its form—say from potential energy to kinetic energy, or vice versa. Other forms of energy can be shown to be either one of these two. For example, current electricity, rotating shaft, moving particles, light, heat, etc., are examples of kinetic energy. On the other hand, static electricity, chemical energy, magnetic energy, energy due to position of a mass in a gravitation field, magnetism, etc., are examples of potential energy.
- Second Law of Thermodynamics: It is impossible for a heat engine to produce a network in a complete cycle if it exchanges heat only with bodies at a single fixed temperature (Kelvin-Planck statement). Also, it is impossible to construct a device which, operating in a cycle, will produce no effect other than the transfer of heat from a cooler to a hotter body (Clausius statement). Second Law thus limits the efficiency of conversion of heat energy into mechanical work, and also gives rise to the concept of entropy. Entropy is a measure of dis-orderliness. It increases with irreversibility. Two main causes of irreversibility in industrial systems, such as that of metal industries, are heat transfer through finite temperature difference and friction. Therefore, no practical system can be reversible, nor attain the theoretical limit of thermodynamic efficiency of reversible processes. Entropy may be considered as the amount of unavailable energy within a given system.

- Third Law of Thermodynamics: It is impossible by any procedure, no matter how idealized, to reduce any system to the absolute zero of temperature in a finite number of operations.
- Dark matter and dark energy: According to our senses and that of our measuring instruments, if no matter nor energy is found in any space, we tend to describe that space as empty. But actually, empty spaces may not empty in absolute sense, as it is estimated that about 30% of matter and 70% of energy of the universe is yet out of our identification/apprehension, even with the help of most advanced gadgets and sensory systems.
- Heat engine: A device, operating in a cyclic process, used to deliver mechanical work or shaft work by converting only up to a fraction of thermal energy input, generally obtained by burning a fuel at a temperature higher than the environment.
- Carnot cycle: A theoretical cycle of a heat engine, with working fluid being always in a state of ideal gas, operating at maximum thermodynamic efficiency between a source temperature and sink temperature comprising four thermodynamic processes, viz. isothermal heat addition (expansion), reversible adiabatic expansion, isothermal heat rejection (compression) and reversible adiabatic compression.
- Carnot cycle, as represented in Figure 3.3, places before us a theoretical limit to attain the highest possible thermal efficiency of conversion of input heat to output work. It also reveals that the same thermal energy at higher temperature has the higher potential to be converted to mechanical work—so the higher the temperature of thermal source, the higher will be its grade. Electrical energy has the potential to be completely converted into mechanical work, and therefore are the examples of high-grade energy. Heat is low-grade energy; as the temperature of the heat source increases, the level of grade also will increase. That is why the Second Law of thermodynamics is sometimes termed as the law of gradation of energy.

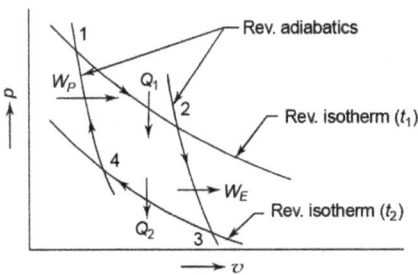

Energy interactions in the Carnot cycle

1–2: reversible isothermal heat addition Q_1 at T_1 ($=T_2$) in Kelvin

2–3: reversible adiabatic expansion, W_E work out

3–4: reversible isothermal heat rejection Q_2 at T_3 ($=T_4$) in Kelvin

4–1: reversible adiabatic compression, W_P work in

Therefore, Carnot efficiency η_C

$= $ (desirable output) / (necessary input)

$= $ (turbine work $-$ pump work)/ $Q_1 = (W_E - W_P)/Q_1$

$= (T_1 - T_3) / T_1 = (t_1 - t_3) / (t_1 + 273)$

$= $ (temperature of source $-$ temperature of sink) / (temperature of source in Kelvin)

FIGURE 3.3 Thermodynamic representation of Carnot cycle on p-v plane and mathematical expression.

- Rankin cycle: A thermodynamic cycle of a heat engine, having the potential to work in practice using fluid changing phase between gaseous and liquid states, operating between a source temperature and sink temperature comprising four thermodynamic processes, viz. isothermal heat addition (in a boiler), reversible adiabatic expansion, isothermal heat rejection and reversible adiabatic compression.

3.3 GOING BEYOND THERMODYNAMIC EFFICIENCY

Gunderson and Yun (2017) pointed to a growing body of literature, partially due to the widespread and common existence of the Jevons paradox, showing economic growth to be antithetical to environmental sustainability and suggesting that economic degrowth and steady-state economic policies could offer alternatives to green growth or win–win mechanisms for the economy and the environment. Modern economics and planning activities therefore deal with EE&C goals in light of longer-term and more holistic gains through various expressions of the term, as explained in Table 3.1a and Table 3.1b.

Besides those in the preceding tables, there are popular terms also used in metal industries in similar spirit that of efficiency, but they are characterized in an opposite manner. Some of these terms are as follows.

Heat rate (HR): Applicable to any thermal system producing electricity. Expressed as the amount of thermal energy input through fuel per unit (kWh) electricity generated (kCal/kWh). It may be noted that thermal efficiency (η_{th}) is reciprocal of this expressed in same unit i.e. 860 / (HR), as 1 kWh = 860 kCal (only unit conversion, not energy conversion).

Specific energy consumption (SEC): Frequently used to track the overall energy performance of metal production system. Expressed as the amount of total energy input (each type of energy input may be converted to any one and only one of kg oil equivalent or GJ, etc., and then summed up) per unit amount of metal produced (GJ/tcs, i.e. gigajoule per ton of crude steel produced). This is reciprocal of energy physical efficiency (e_p).

Energy intensity (EI): In a broader sense, this energy performance indicator is also a kind of SEC. However, EI is generally applied in the context of nations and is expressed as the ratio of total energy consumption expressed in one unit, say, tons of oil equivalent and GDP (toe/USD). The more efficient the economy of a nation, the less would its EI be.

Energy use index (EUI): This is expressed as the ratio of total energy (all forms of energy used are expressed in any one and only one unit, say, kWh) to total air-conditioned space that consumed this energy in a certain period, say, a month or an year, and is expressed as kWh/m².

Energy cost index (ECI): This is expressed as the ratio of total energy cost (expenditure made for all forms of energy used are expressed in any one and only one currency, say, INR) to total air-conditioned space that consumed this energy in a certain period, say, a month or an year, and is expressed as INR/m². ECI takes into account the grades or qualities of various forms of energy used.

TABLE 3.1A

Various Expressions of Energy Efficiency Based on Thermodynamic Aspects (Nag, 2009) i.e. Thermodynamic Energy Efficiency (All Are Dimensionless Numbers, between 0 and 1, Both Inclusive, i.e. $0 \leq \eta \leq 1$)

Sr No	Term (Notation)	Expression	Example of Implication in Metal Industries	Points to Ponder	Tips on Top
1	First Law efficiency *or* energetic efficiency *or* thermal efficiency (η_{th})	(Desired energy output) / (necessary energy input)	Reheat furnace efficiency = (heat transferred to metals reheated during a certain period) / (caloric value of fuel burned in the furnace during the same period).	Continuous casting may reduce/ eliminate the need for a reheating furnace. Also, applications of heat pump may be found to be more effective than efficient furnace in some low temperature heating applications.	Find out more effective rather than more efficient process.
2	Carnot efficiency (η_C)	(Network output) / (necessary heat input) *or* (turbine output-pump input) / (heat input to boiler) *or* (difference in temperature of source and sink) / (temperature of source expressed in Kelvin)	Used as a reference point to compare actual heat engine cycle efficiency (say, that of a captive power plant) with that of theoretical limit of ideal reversible cycle.	Applicable in a reversible heat engine operating in a cycle. For all calculations with temperature, use only the absolute scale. Difference of temperature will be same for centigrade and absolute scale.	At least two practical significance of this impractical efficiency: First—parametric study, e.g. how efficiency can vary with source temperature; Second—how far practical achievement is behind the theoretical limit.
3	Second Law Efficiency *or* exergetic efficiency (η_{II})	(Desired energy output) / (maximum possible energy output)	It is a measure of reversibility of thermodynamic cycles such as a power plant.	Cycles with no irreversibility, such as Carnot cycle, has $\eta_{II} = 1$ or 100%	η_{II} can serve as a tool to check validity of the cycle. If $\eta_{II} > 1$ or $\eta_{II} > \eta_{th}$, the cycle is impossible.

(Continued)

TABLE 3.1A　Continued

Sr No	Term (Notation)	Expression	Example of Implication in Metal Industries	Points to Ponder	Tips on Top
4	Internal or isentropic efficiency (η_i)	(Internal output) / (ideal output for isentropic process)	All pumps, compressors and turbines; for example, blast furnace top gas pressure recovery turbine.	In case of pump and compressor, instead of output, input to be considered.	If not mentioned, or data not available, assume $\eta_i = 1$
5	Mechanical efficiency (η_m)	(Brake output or shaft output) / (internal output)		In reality, it is difficult to measure. A laboratory setup would be necessary.	If not mentioned, or data not available, assume $\eta_{brake} = \eta_m$.
6	Brake efficiency or shaft efficiency (η_{brake})	(Brake output or shaft output) / (ideal output)		Product of internal or isentropic efficiency and mechanical efficiency.	$\eta_{brake} = \eta_m \times \eta_i$
7	Generator or equipment efficiency (η_{gen})	(electricity or equipment output)/(mechanical or equipment input)		In reality, it is difficult to measure. A laboratory setup would be necessary.	Use the chart to estimate η_{gen}.
8	Overall system (incorporating multiple equipment) efficiency (η_o)	Boiler efficiency × turbine efficiency × generator efficiency	All metal industries are likely to have captive power plants. It would be usual practice to calculate η_o.	Calculate value often less than that claimed in technical manual.	Check maintenance and other criteria (such as quality of fuel) are as per guidelines in manual. Otherwise, find the necessary corrective action to be taken.

TABLE 3.1B

Various Expressions of Efficiency (1 ≥ η ≥ 0) Based on Economic Aspects Used along with Energy Efficiency (Belt, 2017; Wei and Liao, 2016) i.e. Economic Energy Efficiency (All Are Numbers with Dimensions and Different Units, Such as Ton/GJ, Ton/$, etc.)

Sr No	Term	Expression	Example of Implication in Metal Industry	Points to Ponder	Tips on Top
1	Energy macro-efficiency (e_m)	(GDP or economic production) / (energy consumption) in, say, $/kgoe Where: In case of country or state GDP to be taken based on PPP In case of metal industry, production value in $	Dimitropoulos (2007) suggested that the importance of the macroeconomic rebound effect on energy indicators should not be underestimated.	Other similar terms are "energy efficiency of economic production" (Tharakan et al., 2001), energy production rate, reciprocal of "energy intensity," etc.	The inverse of this term is known as energy intensity. Improvement in energy macro-efficiency in metal industry indicates most likely proportional increment in productivity.
2	Energy physical efficiency (e_p)	(Production in ton) / (energy input in GJ)	This is inverse of specific energy consumption of metal production (GJ/ton).	Does not differentiate with various grades or costs of input energy forms; only takes its GJ value.	Higher the value of e_p, lower will be its energy intensity and likelihood to be better economically.
3	Energy value efficiency (e_v)	(Production in ton) / (sum of the values of energy inputs in $ or sum of all energy inputs converted to its thermal equivalent in GJ)	This is the inverse of energy cost per ton of production or energy input (GJ) per ton of production.	While converting all energy inputs to its thermal equivalent quantity (say, GJ), appropriate unit conversion and energy conversion to be applied to each input energy quantity.	Calculation to obtain e_v is simple and can be used to track this to check consistency of operation.

(Continued)

TABLE 3.1B   Continued

Sr No	Term	Expression	Example of Implication in Metal Industry	Points to Ponder	Tips on Top
4	Energy allocation efficiency (e_a)	(Optimum elements of input, including energy, of minimum expenditure to company for same output—magnitude of deviation from actual) / (actual input of elements, including energy, for given output)	Metal industry can utilize this tool to maximize e_a by minimizing the input expenditure. e.g. in the iron and steel industry, if, a particular combination of DRI, sintered iron and hematite can reduce production cost when compared to only hematite.	A lot of data and estimates will be required to calculate and update the value of e_a.	It boosts the morale of stakeholders to find the optimum energy use solution. For example, if coal price is reduced, lower grade cheaper ore may be used with more coal to produce metals at higher e_a and e_e.
5	Energy utilization efficiency (e_u)	(Maximum output obtained from same quality and quantity elements of input by optimization) / (actual output for given elements of input including energy)	Operation and process optimization helps to improve e_u, and hence higher productivity with same resources.	"Total Factor Energy Efficiency" may be a better name.	Data envelopment analysis (DEA) may be used to find e_u.
6	Energy economic efficiency (e_e)	Energy allocation efficiency × energy utilization efficiency = $(e_a) \times (e_u)$	Metal industries may use this tool to enhance profitability by effective resources planning, process and operation optimization.	Understanding of thermodynamics, process, statistics, economics and costing are needed to use this tool in metal industries.	Dedicated continuous effort is necessary in metal industries to monitor relevant data on a continuous basis, say, by energy manager.

3.4 MOTOR SYSTEMS

Electric motors are very vital in industrial applications, as their applications are applied in compressor fans, pumps, drives and blowers, which are used in industrial plants. Globally, it is proved that 60% of electrical equivalent to 405 MTOE are consumed in industries. The International Energy Agency calculates significant energy saving of 140–190 MTOE by implementation of new technology and up gradation [3]. There are several methodologies and techniques in which huge amount of energy savings in motor driven systems can be attained. These are described in what follows.

3.4.1 EXISTING MOTORS

Motors should always be switched off when they are not in use; when these motors are aligned and concealed with other accessories and equipment and they are mismanaged as far as operational point of view is concerned and switched on when they are not in use, this leads to wastage of energy and cost. Such situations must be identified and monitored strictly by the energy audit team periodically, and workers should be technically trained for specific controls of motors. In order to mitigate the efficiency and effectiveness of existing motors, they should be tested and should be studied in depth, and it may be possible to redesign a manufacturing process to reduce the use of the motor or to make best use of its effectiveness when it is running state.

3.4.2 MOTOR ALIGNMENT

Pump and motor alignment are the case studies which are to be looked upon in depth of energy inefficiency. A misaligned motor and pump can lead to wear and tear failure. Previous literature findings about electric pumps have proven results that aligning a misaligned motor correctly will give significant better results with power savings of 2.3%, while losses are increased to 8–9% for misaligned motors.

3.4.3 SIZING OF MOTORS

The oversizing of pumps and motors leads to depletion of energy, and this leads to unnecessary financial penalties in terms of costs like energy and maintenance costs. Oversizing can happen due to a number of reasons. The motor and pumps suppliers are of the opinion for procurement of the higher rating which will meet flow requirements, which results in increasing the capital costs; suppliers should be properly chosen and the issue of safety margin should be properly investigated. The substitution of motors with higher modification will lead to increases in their operating capacity.

3.4.4 USE THE MOST EFFICIENT MOTOR

The specifications—that is, size and rating of the motor—determine the loading and overloading of the motors, and efficiency of the motor changes on the basis of loading of the motor; therefore, correct size of motors are to be selected with proper design and safety margin and loading conditions. Running cost of the motors is 1,000 times greater than the actual cost of the motor. Therefore, to operate a motor with energy-efficient objective is of prime importance for industrial applications.

Table 3.2. shows classification of motors according to IEC; as a result IE3 motor in continuous use will use almost 4% less energy per year than an equivalent IE1 model.

Energy-efficient motors operate with efficiencies that are 3–4% higher than standard motors (Figure 3.4). Bureau of Indian standards (BIS) energy-efficient motors are designed to operate without loss in efficiency at loads between 75% and 100% of rated capacity. This may provide overall benefits in transient load conditions. The power factor is about the same as—or may be higher than—that for standard motors. Some countries are banning use of rewinding motors and establishing mandatory requirements with minimum energy performance standards (MEPS) for higher efficient motors.

TABLE 3.2
IEC Classification of Motors

Label	Energy Efficiency	Notes
IE1	Standard efficiency	Efficiency levels to the existing European CEMEP efficiency 2 class
IE2	High efficiency	Same as the minimum efficiency levels set by the US Energy Policy Act (EP Act)
IE3	Premium efficiency	Identical to NEMA Premium in the United States for 60 Hz
IE4	Super premium efficiency	Higher than any other standard; it is not yet implemented

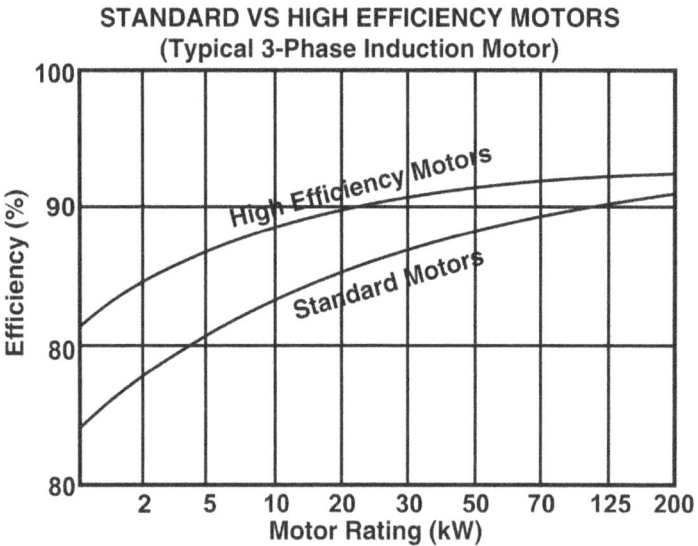

FIGURE 3.4 Comparison of energy-efficient motor with standard motors.

3.4.5 MAINTENANCE ISSUES

Electrical motors should be periodically checked, including preventive and break-down maintenance which include pulleys alignment and checking conditions of belts and tension, lubrication and loose terminals. Second, it should be checked if the supply voltage is within the specified limit or not from the motor's rated voltage (+/− 5%) and the balancing of line voltage. The most common type of motor is an AC induction motor.

3.4.6 VARIABLE-SPEED DRIVES

Variable-Speed Drives (VSDs) are used to vary speed of a motor to its maximum rated speed to match the actual power demand of the load, and perfectly designed VSD systems can reduce significant energy consumption. The payback period of retrofit VSD is of less than 1 year. It is a proven fact that 50% of industries and their processes would benefit from VSDs [2, 3 and 4]. The significant use of VSDs will give huge energy savings to all countries. The total saving of EUR 56,404 million [5] was reported if the full potential of VSDs was implemented across ten countries, (United States, China, Russia, Germany, India, United Kingdom, Spain, France, Turkey and Poland),

3.4.7 MOTOR CHARACTERISTICS

Motor speed is defined as the number of revolutions per minute (RPMs). The synchronous speed of an AC motor depends on the frequency and the number of poles of the motor. The synchronous speed in RPMs is given by the following equation.

$$N_s = 120f / P$$

Where f is frequency and P is the number of poles of motor.

The actual speed will be less than the synchronous speed. The difference between synchronous and full load speed is called slip, and is measured in percent which is as follows.

$$S\% = (N_s - N_r) / N_s$$

Where
N_s = Synchronous speed
N_r = Rotor speed

3.4.8 POWER FACTOR

Power factor is generally represented by the following equation.

$$\cos\phi = kW / kVA$$

Where kW is in kilo Watt and kVA is kilo volt ampere

If the motors are underloaded, the active current is reduced and henceforth there is no reduction in the magnetizing current, which is proportional to supply voltage, resulting in low power factor. Induction motors, operating in under load conditions, are the main cause for low power factor in electric systems.

3.4.9 Motor Efficiency and Field Test for Determining Efficiency

Two significant terms for motors are efficiency and power factor which are used in industries. Motor efficiency (η) is defined as the ratio of the mechanical power developed at the rotating shaft to the electrical input, and power factor (PF)—depending on the type of loads—is more inductive in nature for induction motors. If PF less than 1, resistance losses will be higher, as these are proportional to the square of the current. Thus, both a high value for η and a PF close to unity are desired for efficient overall operation in a plant. Squirrel cage motors are normally more efficient than slip-ring motors.

3.4.10 No Load Test

Figure 3.5 gives overall performance of motor with respect to power factor, efficiency and percent of rated load. It has been experimentally observed for lower percentages of loading power factor and efficiency gives rising characteristics. But as soon as the percentage loading increases to 50–60%, the efficiency curve becomes smooth and stable and the motor runs smoothly and with less friction and windage losses, although power factor shows a flattening of curve with less rising characteristics at 75–95% of loading. So, for the best operating characteristics, it is experimentally

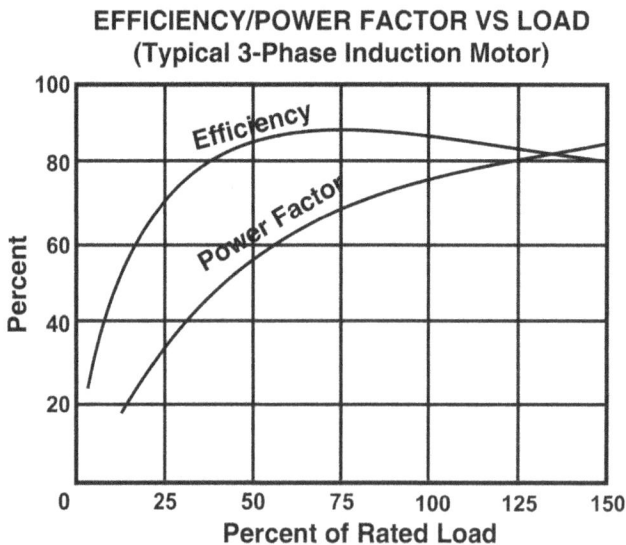

FIGURE 3.5 Percentage of load vs. power factor efficiency [2].

FULL-LOAD POWER FACTORS AT VARIOUS SPEEDS
(Typical for 50 hp Squirrel-Cage Induction Motors)

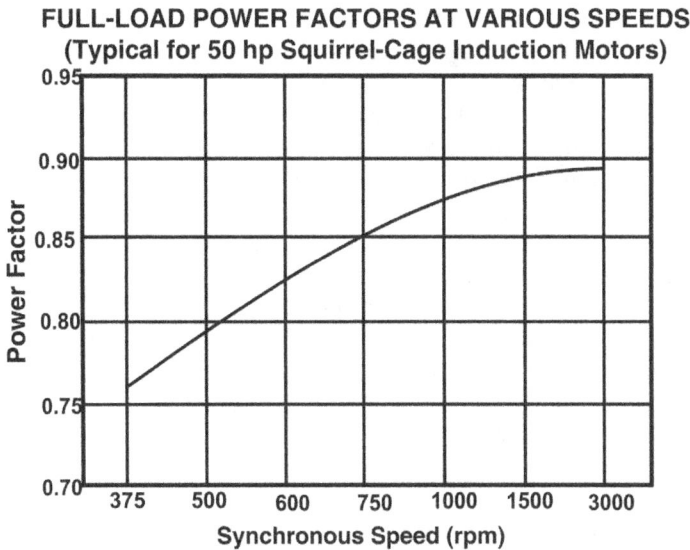

FIGURE 3.6   Speed vs. power factor.

proven that motor loading gives best performance at 60–70%, depending on operating conditions.

F&W and core losses = no load power (watts) – (no load current)^2 × stator resistance

Figure 3.6 gives overall performance for full load power factors at various speeds, specifically for 50 hp squirrel cage induction motors. At lower synchronous speeds of 375–600 rpm, the power factor is lower of the order of 0.77–0.82, but at higher speed of the order at 1,000–3,000 rpm, the power factor becomes flattened and smooth, and is of order of 0.88–0.90.

3.5   ELECTRIC FURNACE SYSTEMS

Blast furnaces and electric arc furnaces are the two types of furnaces which are used in the iron and steel industry. Blast furnaces are extensively used for rotor steel making and steel roll ballets. Electric furnaces are of mainly of two types: arc furnaces and heating furnaces. The energy conservation measures and potential for the iron and steel are described in what follows.

1. Scrap steel pretreatment and processing: scrap steel of tiny and smaller size is first treated, and the overall advantages of pretreatment are that it increases overall heat efficiency and that the electricity consumption per unit of production is reduced.

2. Feeding times reduction of scrap steel: radiation heat loss from the body of the furnace is reduced, and feeding times reduction of scrap steel is also significantly reduced.
3. Control of electricity and optimization: electric current must be controlled at the start of the charging process to maintain electric arc stability and for protection of the furnace cover. An ample amount of current is required for the steel melting process to be faster during the process of primary melting.
4. Reduction period is to be shortened: limestone and quicklime is used; small particle size can accelerate the reduction.
5. Oxidation and melting period reduction: an oil oxygen combustor is used for the reduction of electricity consumption, simultaneously increasing productivity and reducing electrode consumption.
6. Reduction of production costs by using large arc furnace.
7. Water-cooled furnace cover and wall can maximize electricity by using water-cooled furnace at the time of smelting which increases the life of the furnace cover and wall.
8. Energy conservation and energy-saving opportunities in oxygen and carbon blowing.

3.5.1 Dust Vacuum Equipment

Direct and indirect sucking techniques are used for dust collection, and it must increase its range of collection of dust with instant diffusion of rising dust from the arc furnace. The process results in a large volume of air, and plenty of costs are incurred for dust collection equipment. Advanced dust collection equipment has scrap steel pre-heating and air blower rotation control, which results in significant energy savings—0.7 kg/ton electrode consumption saving—and 10% smelting time reduction, while using rotation speed control variable voltage variable frequency (VVVF) for the air blower can achieve significant energy savings.

3.5.2 Continuous Foundry Equipment

The benefits of continuous foundry equipment include low energy consumption, high efficiency and quality. Continuous foundry is globally used by the steel industry due to stability and controllable nature, and the steel bloom and billets produced are uniform, dense and little segregated. The continuous foundry process achieves further savings in processing thermal energy (with fuel oil savings of approximately 30%).

3.5.3 Heating Furnace

Recuperation is one of the most significant energy-saving measures, and when the pre-heating air temperature is from 200–600°C, energy savings of 10–35% are reported from different case studies. The two main indicators for high capacity recuperator (HCR) performance are HCR rate and steel feed temperature. HCR rate varies with production condition between 20% and 80%. When steel feed temperature is 300–850°C and HCR rate is high, the benefits of energy savings are also high.

1. Combustion air ratio. Fuel utilization efficiency is high in ideal conditions and to support complete combustion, it is recommended to adjust air-to-fuel (A/F) ratio according to level of excess air for removal of steel oxide (scale) to reduce acid stripping time and avoid local overheating owing to high-temperature flame.

2. Furnace temperature control. There are three types of furnace temperature control sections: pre-heating section, heating section and uniform heating section. In each section, control of temperature is required, and when furnace temperature is high, combustion gas temperature is also high and heat loss is significant. Reducing furnace temperature as and when required helps increase combustion efficiency.

3. Furnace pressure. There is always opening and closing of doors in an ordinary heating furnace for the bloom and billets entering and leaving. There is significant heat loss if the cold air from outside enters and if there is negative pressure. The furnace pressure should be set slightly higher than the normal atmospheric pressure. Positive furnace pressure causes less heat loss than negative pressure. Henceforth, it is better for furnace pressure to be positive than negative.

4. Addition of ceramic fiber. The following effects are observed with the addition of ceramic fiber.

    i. The heat loss from the furnace body is reduced significantly with the addition of ceramic fiber—as a result increasing the heating rate. As a result, operating temperature reaches very quickly and thus increase work time.

    ii. According to the previous studies, the benefits are not significant for HCR and longer furnace, and therefore it is highly recommended for feasibility investigation for financial gains is suggested before ceramic fiber patching. Numerous furnaces suffer peeling problems with ceramic fiber patching, and their energy-saving benefits are not achieved.

    iii. Loss from furnace opening. The pressure difference between the inside and outside of the furnace is positive, and the lower the pressure difference, the better the heat conservation.

    iv. Reduction of loss from opening radiation. Radiation heat loss is determined by temperature, opening size and radiation coefficient. Temperature is essential to furnace operation. Thus, the best method of minimizing radiation loss is to reduce the opening size or even close the opening completely.

    v. Reheating furnace (withdraw type). If the output mode of the reheating furnace is pusher type, cold air frequently results from the uniform heating section and produces high oxygen content. It is best if the output mode is the withdraw type, which improves oxygen uniformity and reduces scale loss to 1.5%.

5. Air pre-heating. This measure can generally achieve energy recovery exceeding 10%, and thus is one of the most effective energy savings measures. Energy savings of 25–35% are achieved when air pre-heating temperature is 600°C. Meanwhile, air pre-heating temperature of 200°C achieves energy savings of 10–20%.

3.6   FURNACE SYSTEMS

Furnaces are a region in space, usually closed, where chemical energy of fuel (e.g. coal, oil, biomass or gas) or electrical energy (i.e. alternating current [AC] or direct current [DC]) is converted to thermal energy or heat, which is transferred to certain materials inside the furnace in order to either melt (e.g. melting furnace, blast furnace, arc furnace, induction furnaces, etc.) or boil (e.g. boiler, coal-based thermal power plant), or heat treat (e.g. heat treatment furnace, annealing furnace), or only to raise temperature (e.g. reheat furnace). Almost all the previously mentioned types of furnaces are indispensable in metal industries, both in the metal extraction process and for captive power generation. Improvement of EE&C in furnaces of metal industries is a continuous process for economy of production, comfort in work environment, complying with the pollution control norms, enhanced life of the furnace system, etc.

As introduced in Chapter 1, energy efficiency and energy conservation are two correlated but distinct energy-specific terms. In Figure 3.7, across the system boundary of the furnace, there are three energy interactions, viz. fuel input, heat transferred to metal (in the form of heated metal job output taking metal job at room temperature as input) and system losses. According to the First Law, input heat (mass of fuel × caloric value) is equal to heat transferred to metal (mass of metal job × specific heat × temperature rise as it moves from input to output) plus system losses (heat in flue gas + heat in unburned fuel + radiation, convection and other losses), as shown in following equation for any period of time (say, one hour or day or week or month or even year).

$$m_f \times GCV = m_j \times s_j \times (t_o - t_i) + \Sigma Losses \qquad (3.1a)$$
$$m_j \times s_j \times (t_o - t_i) = m_f \times GCV - \Sigma Losses \qquad (3.1b)$$

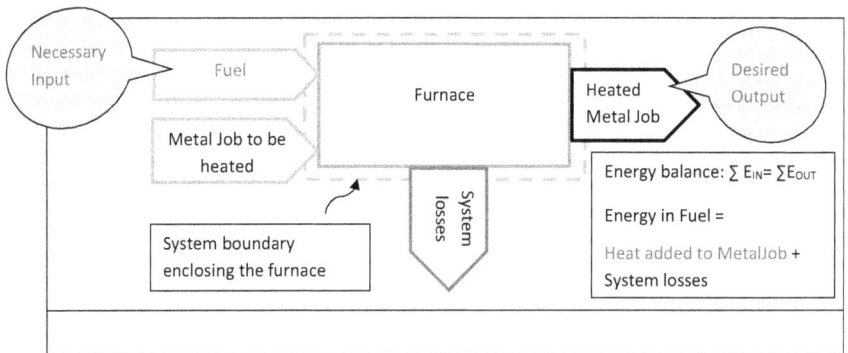

FIGURE 3.7   Energy balance of a furnace system.

In most cases, flue gas temperature at furnace outlet is more than 100°C.

Σlosses

= Heat taken away by moisture of fuel and moisture formed during combustion due to hydrogen in fuel + Σall other losses

= 584 kCal/kg × m_j (%inherent moisture in fuel) + 9 × m_j (%hydrogen present in fuel)} + Σall other losses

Thus, Eqn. (3.1a) becomes,

m_f × GCV = m_j × s_j × $(t_o - t_i)$ + 584 kCal/kg × m_j (%inherent moisture in fuel) + 9 × m_j (%hydrogen present in fuel)} + Σall other losses

i.e.

m_f × GCV − [584 kCal/kg × m_j (%inherent moisture in fuel) + 9 × m_j (%hydrogen present in fuel)}]= m_j × s_j × $(t_o - t_i)$ + Σall other losses

i.e.

m_f × [GCV − 584 kCal/kg × (% inherent moisture in fuel) + 9 × (% hydrogen present in fuel)] = m_j × s_j × $(t_o - t_i)$ + Σall other losses

i.e.

$$m_f \times NCV = m_j \times s_j \times (t_o - t_i) + \text{Σall other losses} \qquad (3.1c)$$
$$m_j \times s_j \times (t_o - t_i) = m_f \times NCV - \text{Σall other losses} \qquad (3.1d)$$

Where GCV (gross calorific value of the fuel, also known as high heat value [HHV]) is the amount of heat released by complete combustion of one kg of fuel in air, and

NCV (net calorific value, also known as lower heat value [LHV]) is the amount of heat released by complete combustion of 1 kg of fuel in air. less the heat taken away by the moisture present in fuel and moisture formed due to hydrogen present in fuel.

Here, instead of latent heat of evaporation of water (540 kCal/kg), a statistical average value of 584 kCal/kg is taken to take in to consideration a slightly negative pressure in furnace and specific heat of steam being lower than that of water, as explained in Chapter 3.

Now the efficiency of the furnace may be expressed from Equations 3.1a and 3.1c, as follows.

$$\eta = m_j \times s_j \times (t_o - t_i) / (m_f \times GCV) \qquad (3.2a)$$

But, it would be more realistic to consider NCV as in general flue gas is left at more than 100°C to protect the furnace system from erosion by acid which is formed in flue gas at a temperature lower than 100°C—and therefore, the furnace should not be hold responsible (by lower value of η with the use of GCV), as follows.

$$\eta = m_j \times s_j \times (t_o - t_i) / (m_f \times NCV) \tag{3.2b}$$

Efficiency calculations expressed in Eqns. 3.2a and b are called direct method. The efficiency can also be expressed from Eqns. 3.1b and d as follows.

$$\eta = (m_f \times GCV - \Sigma losses) / (m_f \times GCV) \tag{3.2c}$$
$$\eta = (m_f \times NCV - \Sigma all\ other\ losses) / (m_f \times NCV) \tag{3.2d}$$

Numerators in Equations 3.2c and 3.2d are same, but the denominators are different. This approach of calculating the efficiency is known as indirect method or loss calculation method. As the name suggests, this method helps us to identify significant losses separately and thereby justify its value to be higher or lower than the standard or baseline—if it is more, there is opportunity for improvement; if it is less, there if opportunity for replication.

However, it is a normal practice to first calculate the efficiency using direct method, as it is easy and necessary to first compare it with standard or baseline efficiency value. If it is found to be less, further investigation becomes possible by employing the indirect method.

This approach of calculating efficiency by direct and indirect method is applicable to electrical systems, as well. For audit-related discussion, please refer to Chapter 5.

3.7 EE&C IN (CHEMICAL) UNIT OPERATIONS

Metals are extracted from ores available in nature, but their reclamation, concentration and processing for extraction of metals is an energy-intensive process. For reasons explained in Chapter 1, the manufacturing process of iron, steel and aluminum are the focus area of discussions in this book. General conversion factors from metal content toward gross ore, and ore concentrate toward gross ore "run-of-mine" are shown in Table 3.3. The corresponding values for iron and aluminum are 43% and 82%, and 19% and 68%, respectively, which are among the highest when compared with other metals in the table. These are the two metals more extensively used today (followed by copper) and share maximum of the energy consumption in metal industry sector. The metal extraction process from the ore involves a series of stages in which there is scope for continuous improvement in terms of energy intensity, productivity and quality.

EE&C opportunities in such metal manufacturing processes are elaborated in Chapter 4. This section is a linkage of the previous sections with Chapter 5. In previous sections, the EE&C of various gadgets and equipment used in the metal industry—such as furnaces, boilers, motors, pumps, compressors, blowers, etc.—have been discussed. In this section, the various unit operations of metal manufacturing, comprising those gadgets and equipment, will be discussed. This is necessary because the objective and practice of EE&C of individual equipment may appear to be sometimes opposite to that of unit operations comprising that equipment. For example, in a blast furnace, with higher loading, the SEC may reduce because of

TABLE 3.3

General Conversion Factors from Metal Content toward Gross Ore, and Ore Concentrate toward Gross Ore "Run-of-Mine"

Metal	% of pure metal content in gross ore (to convert from 'metal content' towards 'gross ore run of mine')	% of concentrate content in gross ore (to convert from 'ore concentrate' towards gross ore 'run-of-mine')
Iron	43.32	81.93
Copper	1.04	3.33
Nickel	1.83	23.45
Lead	11.86	16.52
Zinc	8.34	14.5
Tin	0.24	0.33
Gold	0.00021	0.0663
Aluminium	18.98	67.55
Silver	0.034	2.552
Uranium	0.0015	0.3744

Source: European Union (2018)

Notes: When specific national conversion factors are not available, general conversion factors may be applied as a default setting; the general conversion factors provided in this table are predominantly based on annual business reports for about 160 metal mines in the EU, which has warned that these data should be used only as a last option because they do not take into account variations across mines and time.

decremented contribution of energy consumption at higher loading. But the furnace cooling system may use disproportionately more cooling water circulation in order to maintain the life of the furnace, and thus, incrementally contribute to the SEC.

3.7.1 EE&C in (Chemical) Unit Operations in I&S Industries

The major unit operations prevalent in the iron and steel (I&S) industry typically involve oxygen plants, coke-oven plants, sponge iron plants, captive power plants, blast furnace sections, electric arc furnace sections, basic oxygen furnace sections, hot metal sections, hot rolling mills, galvanizing sections, cold rolling mills and ferro-alloy sections, as depicted in Figure 3.8. The various metals manufacturing processes are discussed with a flowchart in Chapter 4.

3.7.1.1 EE&C in (Chemical) Unit Operations in I&S Industries: Blast Furnace

Typical features of a blast furnace are:

- Capacity: 2 million tons per annum (TPA)
- Volume: 2581 M^3

Manufacturing Process for Iron and Steel

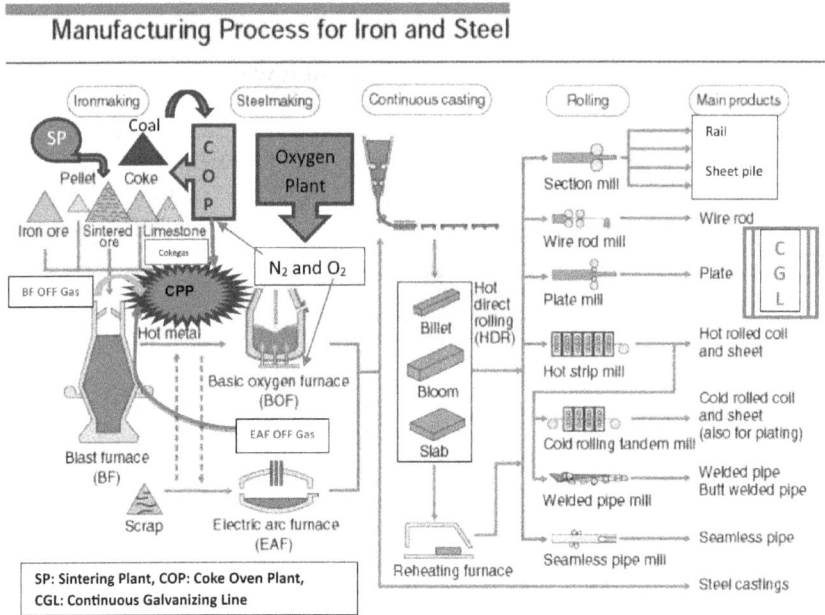

FIGURE 3.8 Various unit operations of a typical integrated steel plant.

Source: www.jfe-21st-cf.or.jp/chapter_2/index.html

- Twin tap hole/twin cast house with slag granulation plant
- State-of-the-art integrated Level II automation
- Top pressure recovery gas expansion turbine

Brief description of the process: blast furnace is a process to convert the iron ore to iron by the process of reduction. The iron oxides in the iron ore are reduced to iron by the carbon monoxide, which acts as a reducing agent. This carbon monoxide is obtained by burning of the coke.

The major inputs to the blast furnace are:

- Iron-bearing material (iron oxides) such as ore, pellets and sinters
- Coke
- Air required for combustion of coke
- Fluxes

The major outputs from the blast furnace are:

- Liquid metal containing approximately 93% Fe
- Slag generated from the impurities in the raw material
- Blast furnace gas generated by the combustion of coke and other reduction processes, consisting of 20–23% carbon monoxide

A blast furnace is a continuous process plant, i.e., the raw material is fed from the top which goes on getting reduced and finally comes out as liquid metal. This process cannot be stopped suddenly; otherwise, the various metallurgical and thermal equilibriums inside the furnace get disturbed. This differentiates the blast furnace plant from other industries like machine shops, forge shops or other engineering industries.

Electrical power, propane, coal, fine coke and ordinary coke are used in different stages of the production process.

3.7.1.2 EE&C in (Chemical) Unit Operations in I&S Industries: Sponge Iron Plant

Typical features of a sponge iron plant (Figure 3.9) are as follows:

- Capacity: 1.4 MTPA with fully computerized process control
- One of the most energy-efficient plants in the world is gas-based mega-module Midrex plant
- Only plant in the world using lump ore in excess of 60% in feed mix
- Automated lime coating system for high-temperature operation to enhance productivity

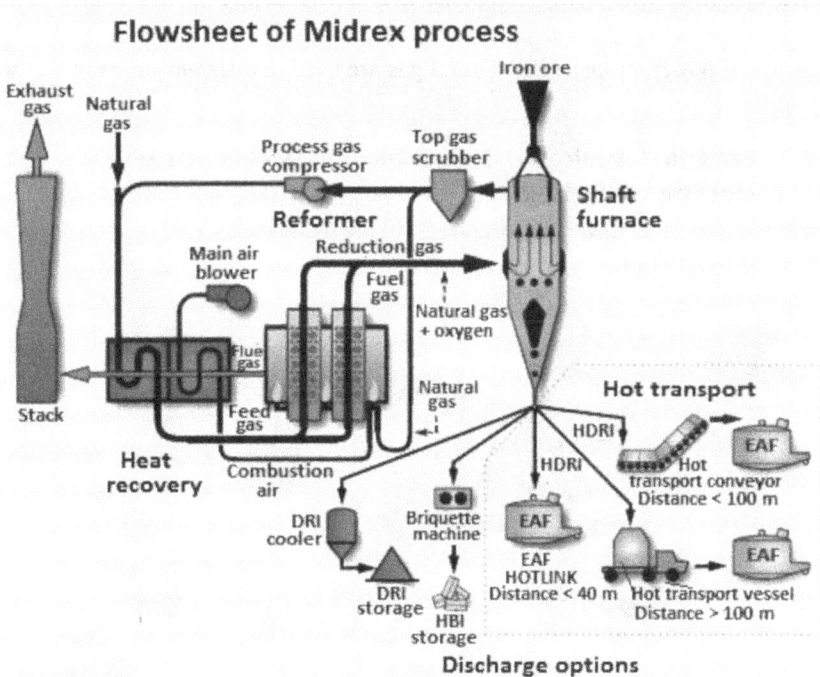

FIGURE 3.9 Sponge iron plant process flow diagram.

Source: www.ispatguru.com/

- Oxygen injection system to increase the bustle gas temperature to enhance productivity
- Process improved consistently to use a high percentage of lump ore in the feed mix to reduce cost of production

Brief description of the sponge iron manufacturing process: it is the process DRI is the chemical reaction of hydrogen and carbon monoxide with iron oxide to produce metallic iron.

It consists of the following processes:

Reduction: iron oxide, as oxide pellets or lump ore, contains about 30% oxygen by weight. In direct reduction, the oxygen in iron oxide reacts with CO and H_2 at elevated temperatures to form metallic iron, CO_2 and H_2O (vapor). The CO and H_2 reduce iron oxide and are known as reductants. The CO_2 and H_2O oxide iron are known as oxidants.

Reforming: the reducing gas is produced from the recirculation of the gas taken from the top of the reduction furnace. This gas is first cleaned by the top gas scrubber. It is then compressed, mixed with natural gas and passed through catalyst-filled tubes. These tubes are heated in a refractory line furnace called a reformer.

Carburization: carburizing is the process by which carbon content of the material is increased. This carbon in the reduced product is essential to the most efficient use of the reduced product in iron or steel making.

Electrical power, propane and natural gas are used in different stages of the production process.

3.7.1.3 EE&C in (Chemical) Unit Operations in I&S Industries: Hot Strip Mill (HSM)

Typical features of a HSM plant (Figure 3.10) are as follows:

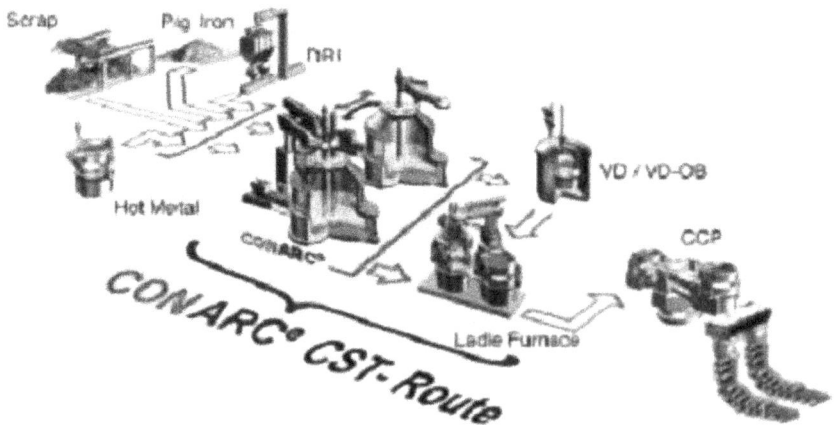

FIGURE 3.10 A typical hot strip mill (HSM) process flow diagram.

- Capacity: 3.0 MTPA
  - Phase I: 1.5 MTPA
  - Phase II A: 0.9 MTPA
  - Phase II B: 0.6 MTPA
- Twin shell CONARC furnace capable of using different combinations of charge mix of hot metal, sponge iron and scrap
- Ability to roll as thin as 1.2 mm with 1,250 mm width
- Capable of producing ultra-low-carbon steels/electrical steels through VOD/VCD/VD route
- HR as a substitute for CR in certain applications due to superior surface quality and thinner gauges
- Rolling mills have HGC, AGC, CVC and roll bending in all six stands
- Roll shifting facility in all stands of +/− 100mm for better profile

Brief description of the process: In HSM, HR coil is produced from the liquid metal. The main raw materials for this section are hot metal from a blast furnace, sponge iron from SIP and electrical power from MSEB. Basically, it is a continuous casting process. Continuous casting is the process whereby molten steel is solidified into a "semi-finished" billet, bloom or slab for subsequent rolling in the finishing mills.

Steel from the electric arc furnace is tapped into a ladle and taken to the continuous casting machine. The ladle is raised onto a turret that rotates the ladle into the casting position above the tundish. Liquid steel flows out of the ladle into the tundish and then into a water-cooled copper mold. Solidification begins in the mold, and continues through the first zone and strand guide. In this configuration, the strand is straightened, torch-cut then discharged for intermediate storage or hot charged for finished rolling.

Next in the process is the tunnel furnace. The roller hearth furnace receives cut-to-length slabs from the caster at casting speed. Both the furnace operational speed and the process heating control is speed-matched to the compact strip production (CSP) caster and thin slab temperature, as monitored at the entrance to the furnace. Length of the soaking furnace plant dimension is such that a roll change can be done during casting also in the second stage. The buffer capacity for the first stage is 20–35 minutes depending on casting speed and slab advance at the beginning of a possible interruption in mill area. Under normal steady operational conditions, the roller hearth furnace is required only to equalize the thin slab's temperature profile, since its inlet temperature of approximately 1,100°C (only at surface) is what is required for discharge into the mill. At the discharge end of furnace, thin slabs have a temperature differential of ± 10°C only along the slab's length, thickness and width. The excellent temperature profile of thin slab entering the CSP mills provides optimum conditions for the subsequent rolling process and thus ensures an excellent profile and flatness of the hot rolled product.

Electrical power, propane and Light Diesel Oil (LDO) are used in different stages of the production process.

3.7.2 EE&C IN (CHEMICAL) UNIT OPERATIONS IN THE ALUMINUM INDUSTRY

Aluminum is manufactured mainly from its principal ore, bauxite, involving extremely energy-intensive processes—60–70% of the production cost goes to

TABLE 3.4

Some Statistics for Captive Renewable Power Plants in Metal Processing Industries in India for the Year 2018–2019 (as of 20 August 2020)

Sr #	Type of Captive Renewable Power Plants	Number of Industries	Installed Capacity (kW)			
			I&S	Aluminum	Others	Total
1	Hydro	5	2550	0	70,916	73,466
2	Solar	108	19,620	13,970	570,950	604,540
3	Wind	73	135,086	50,550	536,907	722,543
4	Total	186	157,256	64,520	1,178,773	1,400,549

Note: Besides these, there are several independent (renewable) power producers (IPPs) which are attached to metal industries.

Sources: Websites of Central Electricity Authority (CEA) of India, Jindal Aluminum Limited and Tata Power Solar

produce or purchase energy. This has been elaborated in Chapter 4. Most of the aluminum industries, like iron and steel, have captive power plants to safeguard the unit operations and also to reduce the electricity generation costs by making maximum use of waste heat available in the production process. Aluminum and copper industries are particularly suitable for renewable power generation, as well, if the location has the potential, particularly solar photovoltaic, because the electrolysis process of purification requires very high–ampere DC and the process can be optimized according to generation, depending on the availability of solar radiation.

3.7.3 EE&C IN (CHEMICAL) UNIT OPERATIONS IN THE ALUMINUM INDUSTRY: CAPTIVE RENEWABLE POWER PLANTS

The first independent power producer (IPP) company to start captive renewable power plant of 4.14 MW wind in India was Jindal Aluminum Limited (JAL), Chitradurga, Karnataka, in 1997. As in 2020, JAL has 50.8 MW wind and 30 MW solar power plants. In India, more than 1,400 MW of captive renewable power plants are operating in metal industries, as shown in Table 3.4.

Case Study 3.1: Jindal Aluminium Limited (JAL) Renewable Captive Power Plants

After successful commissioning of captive wind power plant in 1997, JAL added more wind and solar power plants at Chitradurga and Davangere districts in Karnataka and Anantapur District in Andhra Pradesh, India for the following reasons.

- The aluminum industry—being electricity intensive and having the capability to quickly regulate electrolysis for producing pure aluminum—can better utilize the solar photovoltaic and wind power, though the power production varies with the vagaries of solar radiation and wind velocity.

- Local communities are trained and acquire the necessary skills to maintain and operate the renewable energy systems that earn them their livelihoods and also enhances sustainability of the company.
- Locations, such as Chitradurga in Karnataka, have significant potential for both wind and solar power generation. This, however, called for specially designed solar photovoltaic system that could be seasonally tracked to maximize the power output but sufficiently robust to sustain the wind load. Moreover, the implementing agency, Tata Power Solar, took special care of the systems anchoring to the ground with significant areas covered with water bodies. The 10 MW peak Solar PV system commissioned at Chitradurga in 2013—just four months from the date of land acquisition—did not interfere with the 18-acre natural water body, part of its site (see Figure 3.11). The 51-acre rocky non-agricultural terrain was thus converted in to a solar carpet with ecological balance of the site. The plant produce more than 18,000 MWh of electricity in every year, which is sufficient to

FIGURE 3.11   Solar PV and wind captive power plants both operating at Jindal Aluminum Limited, Chitradurga, Karnataka.

meet the electricity needs of about 70,000 homes in the locality, besides satisfying JAL's captive energy needs. Further, the plant helps JAL also to earn a Renewable Energy Certificate. Another landmark of this project was its capacity, the largest in Karnataka state at the time of its commissioning.

• Such success stories, in both solar and wind captive power plants, have been encouraging more industries to go renewable in an increasing rate now. The government of India has set target of implementation of 100 GW peak solar photovoltaic (SPV) power plant by 2022 and 450 GWe of renewable energy by 2030.

3.7.4 EE&C in (Chemical) Unit Operations in the Copper Industry

Two of the greatest challenges of copper manufacturing from its ore is the dwindling concentration of copper in its ore, and the increasing requirement of 99.9% pure copper for conductors, contactors and energy-efficient gadgets which are increasing in leaps and bounds. McKinsey Global Institute predicts that copper consumption will rise by 43% by 2035, as copper is one of the most important metals for innovation and energy transition (McKinsey, 2017). A breakdown of various usage of copper is shown in Figure 3.12.

Copper is present on an average 33 gm in every ton of Earth's crust collected from top of earth's surface till a depth of 10 km. The copper ore contains a large amount

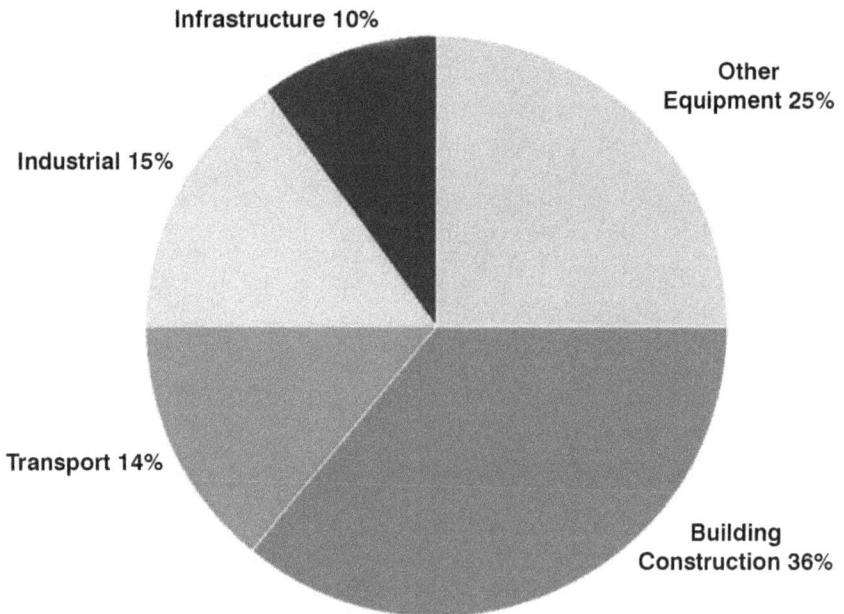

FIGURE 3.12   Approximate breakdown of various usage of copper.

Source: https://copperalliance.eu/

of impurities including clay, dirt and a variety of non-copper materials that reduces the concentration of copper in its ore to less than 1%. Concentration as low as 0.2% is also considered to be suitable for exploration. Both surface and underground mining are in practice, though the former is having larger share (90%) when compared to latter. One important operation, therefore, before the copper manufacturing process can begin, is the concentrating of ore to raise the percentage to more than 20%, typically about 30%. Copper extraction process from concentrate is conducted, similar to that of aluminum, in two steps, viz. smelting and electrolysis.

This unit operation is generally done near mines and then transported to the copper manufacturing facility. The concentrating process involves a froth flotation process which involves mixing pine oil with the slurry of ground ore. The unwanted materials precipitate and can thus be removed. The concentrate can contain other precious elements, as well, such as molybdenum and gold. The step-by-step process from mining ore to production of 99.99% pure copper is shown in Figure 3.13. The manufacturing process is discussed in detail in Chapter 4. Presently, about 50% copper is produced from recycled copper and this should be further increased to improve EE&C and attain circular economy.

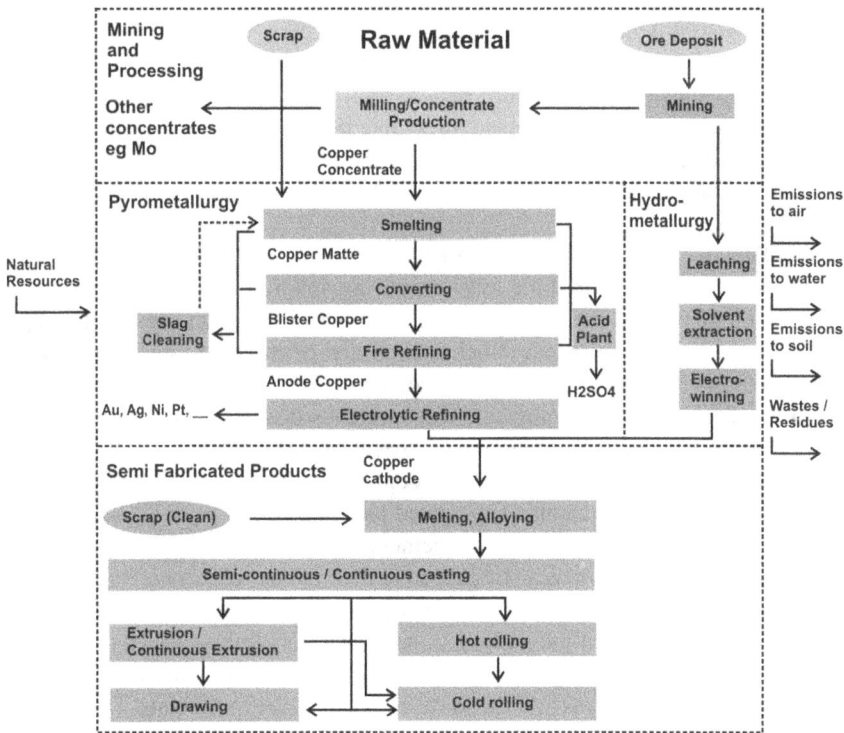

FIGURE 3.13 Flow chart showing the process of mining to manufacturing pure copper and its products.

3.8 EE&C IN WHR, CHP AND OTHER UTILITIES

WHR devices locally act like an energy conservation initiative but improve the efficiency of the system of which it is a part. For example, if a furnace has been releasing the flue gas to environment at say 500°C, it would most likely found to be a techno commercial viable option to employ an air pre-heater to conserve part of the heat loss through flue gas. After deployment of this WHR device, the furnace system efficiency would increase.

CHP devices deliver two industrial utility services, viz. heat and electricity, by energy conversion from a single fuel source. Such systems may be visualized as a successful integration of a WHR device to another energy system which was delivering heat or electricity—the purpose of a WHR device is to deliver also electricity or heat. The direct benefit is an enhancement of efficiency. In some cases, the collateral benefits are multiple, including the following.

1. Lower cost of fuel because of higher efficiency—energy which was wasted earlier is utilized in CHP system.
2. Lower emission because of less fuel burned at improved efficiency and lower thermal pollution—heat rejected from system to surrounding is reduced both in quantity and quality (temperature).
3. Improved work environment for less heat rejection and emission.
4. Improved utilization of space, as CHP will occupy less space than that of two independent systems to deliver heat and electricity.
5. Less manpower requirement. Such system of generation of two services simultaneously, viz. combined heat and power (CHP) from single fuel source (in this case, waste heat only). The authors, while conducting detailed energy audit of large steel plants, identified such opportunities in the following areas.
   i. Sensible heat utilization of off-gas in electric-arc furnace: steam generation is considered the most promising solution to the energy utilization of furnace off-gas during steel making. The steam produced using a waste heat boiler can either be exported or converted into electricity. For efficient utilization of the available thermal energy, the boiler must be designed to accommodate discontinuous operation, gas that is heavily dust-laden and limitations on space for installation. A unique boiler design is proposed for electricity conversion that complies with turbine-operation requirements. Anticipated payback is in 1–4 years, depending on choice of energy utilization method. In highly dynamic processes such as electric arc furnaces (EAFs), steel making the temperature and quality of the off-gas are characterized by widely varying extremes. When the temperature of the gas is low, such as during scrap charging or scrap pre-heating, the off-gas generally contains high concentrations of volatile organic components (VOCs), dioxins and inorganic pollutants. This effect is exaggerated by the ever-increasing percentage of organic combustibles present in the scrap charge, including oily mill scales and organically coated steel, as well as plastics and textiles from automobile recycling.

The conventional solution for neutralizing contaminant-laden off-gas is to install a post-combustion chamber, whereby the VOCs are destroyed at high temperatures (> 800°C). When the content of the organic is less than approximately 100 mg/Nm3, the removal of the dioxins after off-gas cooling is then effectively accomplished through the injection of adsorbents. This means that the energy consumed in the post-combustion chamber is unavailable for other thermal applications. Leading international suppliers of environmental technologies have investigated the main technical and economical factors enabling a fuller utilization of the off-gas energy for steam and possible electrical generation. A detailed description of one such application is explained in Chapter 4.

ii. Application of waste heat steam generation system in tunnel furnace: the roller hearth tunnel furnace receives cut-to-length slabs from the caster at casting speed. Operational speed of both the furnace are synchronous with CSP caster and thin slab temperature. Length of the soaking furnace plant is so dimensioned that for the second stage, a roll change can be done during casting. The buffer capacity for the first stage is 20–35 minutes, depending on casting speed and slab advancement at the beginning of a possible interruption in mill area. Under normal steady operational conditions, the roller hearth furnace is required only to equalize the thin slab's temperature profile, since its inlet temperature of around 1100°C (only at the surface) is required for discharge into the mill. At the discharge end of furnace, thin slabs have a temperature differential of ± 10°C only along the slab's length, thickness and width. The excellent temperature profile of thin slab entering the CSP mill provides optimum conditions for the subsequent rolling process, and thus ensures an excellent profile and flatness of the hot rolled product. The exhaust temperature of tunnel furnace is around 1000°C. After heat rejection at the heat recuperator, the temperature of the flue gas at the chimney base is 675°C average. If a heat recovery boiler can be installed then, steam can be generated which is almost equivalent to the generation of steam by HSM— the boiler, therefore, results in saving of LDO cost. A detailed description of one such application is explained in Chapter 4.

3.9 CONCLUSION

EE&C has been dealt with from fundamentals to frontiers of metal industries. The thermodynamic basis of the concepts further developed in practice have been discussed in this chapter. Economics of process optimization has been exemplified. The unit operation for areas with significance from EE&C point of view has been emphasized. The metal manufacturing process will be discussed in Chapter 4.

REFERENCES

Belt, C.K. (2017). *Energy Management for the Metals Industry* (1st ed.). CRC Press. https://doi.org/10.1201/9781315156392

Dimitropoulos, John (2007); Energy Productivity Improvements and the Rebound Effect: An Overview of the State of Knowledge. *Energy Policy*, Volume 35, Issue 12, December 2007. https://doi.org/10.1016/j.enpol.2007.07.028.

European Union (2018); *Economy-wide Material flow Accounts HANDBOOK*. Luxembourg: Publications Office of the European Union. Accessed on March 2, 2021. https://ec.europa.eu/eurostat/documents/3859598/9117556/KS-GQ-18-006-EN-N.pdf/b621b8ce-2792-47ff-9d10-067d2b8aac4b.

Gunderson, R., and Yun, S.J. (2017); South Korean Green Growth and the Jevons Paradox: An Assessment with Democratic and Degrowth Policy Recommendations. *Journal of Cleaner Production*, Volume 144, pp. 239–247.

Jevons, William Stanley (1865); *The Coal Question; An Inquiry Concerning the Progress of the Nation, and the Probable Exhaustion of Our Coal Mines* (2nd ed.). London: Macmillan & Co. Accessed 2022. https://energyhistory.yale.edu/library-item/w-stanley-jevons-coal-question-1865.

McKinsey. (2017); *Beyond the Supercycle: How Technology Is Reshaping Resources*. McKinsey Global Institute. Accessed in 2022 www.mckinsey.com/.

Nag P. K. (2009); *Basic and Applied Thermodynamics*. Tata McGraw Hill Education Private Limited, 2009. ISBN 1283186993, 9781283186995.

Tharakan, Pradeep J., Kroeger, Timm, and Hall, Charles A.S. (2001); Twenty Five Years of Industrial Development: A Study of Resource Use Rates and Macro-Efficiency Indicators for Five Asian Countries. *Environmental Science & Policy*, Volume 4, Issue 6, pp. 319–332, ISSN 1462–9011. https://doi.org/10.1016/S1462-9011(01)00036-3.

Wei, Yi-Ming, and Hua, Liao (2016); *Energy Economics: Energy Efficiency in China*. Switzerland: Springer International Publishing. DOI 10.1007/978-3-319-44631-8. https://citations.springernature.com/book?doi=10.1007/978-3-319-44631-8.

Wei, Yi-Ming, and Liao, Hua (2016). *Energy Economics: Energy Efficiency in China*. Springer International Publishing. https://doi.org/10.1007/978-3-319-44631-8. U.S. Department of Energy – Energy Efficiency and Renewable Energy.

Websites

[1] https://www.guinnessworldrecords.com/world-records/431420-most-efficient-combined-cycle-power-plant

[2] https://www.npcindia.gov.in/NPC/Files/Publication/Annual-Productivity-Report-2017-18.pdf

[3] https://www.iea.org/reports/energy-efficiency-policy-opportunities-for-electric-motor-driven-systems

[4] https://www.energy.gov/eere/office-energy-efficiency-renewable-energy (U.S. Department of Energy – Energy Efficiency and Renewable Energy)

[5] https://support.industry.siemens.com/tf/ww/en/posts/classes-of-motors/7659

4 Metal Manufacturing Processes and Energy Systems

Swapan Kumar Dutta and Binoy Krishna Choudhury

4.1 GENERIC CHARACTERISTICS OF METALS AND METAL INDUSTRIES

Everybody in the past (for at least the past few thousand years), at present and in future (at least for some hundred years ahead) would accept that some important role is played by metals in our daily life. It is said that humanity has passed through a bronze (an alloy of copper and tin) age (3300–1200 BC), an iron age (1200–600 BC) and an aluminum and steel age (1800–present). Metals have become indispensable in modern civilization for their useful properties such as mechanical strength, electrical conductivity, thermal conductivity, durability, recyclability, etc. (Figure 4.1). More than one metal can be made to form alloys to optimize the desired properties. Metals are generally available in nature in the form of their oxides (or other compounds) from where they can be resourced and are called ores. Modern metal processing techniques to recover metals from their ores and prepare alloys and products for our use strive to manufacture and also to improve the desirable properties of metals (Figure 4.2) in economic, efficient, environmentally benign and sustainable ways. Iron, steel, aluminum, silver and copper manufacturing processes and energy use is discussed in this chapter, as they constitute major applications in our lives (Table 4.1).

4.2 FERROUS AND NON-FERROUS METALLURGY

"Ferrum" in Latin means "metal compounds that contain iron" and includes wrought iron, carbon steel and stainless steel. So, non-ferrous means metal compounds that do not contain iron, such as copper, aluminum, zinc and also precious metals, such as gold and silver. Iron is cheapest of all metals, magnetic and has higher density, higher yield stress and strength, and so has the highest production and use among all metals. Non-ferrous metals are costlier, often lighter than iron, non-magnetic and with higher conductivity. Wrought iron, steel and non-ferrous metals do not rust or corrode, and hence have higher, almost 100%, recyclability for many many years.

Casting yields higher production rates but poor grain structure; hence, mechanical properties as well as durability are better when metal products are wrought. Cast

DOI: 10.1201/9781003157137-4

Pre-painted galvalume sheet (of steel coated with zinc, silicon and aluminum)

Iron and steel for railways and bridges

Silver and alloys, such as bronze and brass, are used in bullion and currency

Transparent compound of aluminum, oxygen and nitrogen (commercially named as ALON)

Copper is extensively used in transformers (as shown), motors, contactors, heat-exchangers, etc.

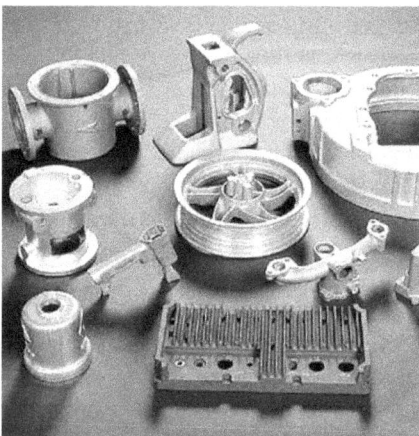

Aluminum and its alloy are indispensable in aviation and automobile industry, utensils, heat-exchangers, etc.

FIGURE 4.1   Various uses of metals in our day-to-day lives.

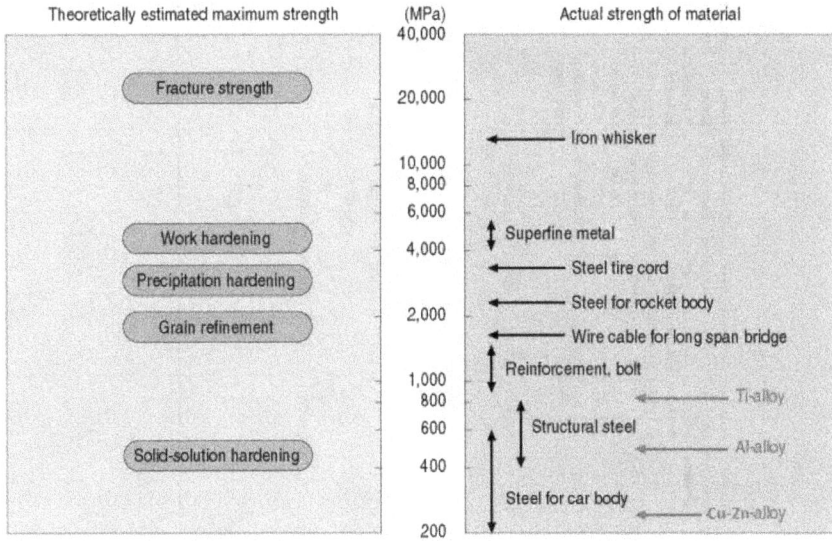

FIGURE 4.2 Theoretical and actual strength of metals and alloys.

alloys are poured molten into sand (sand casting) or high-strength steel (permanent or die casting) molds, where they solidifies to produce the desired shape. This is applicable for both ferrous and non-ferrous metals.

4.2.1 Types of Metals

4.2.1.1 Types of Metals: Ferrous

Ferrous metals, i.e. the iron and steel sector, which represents 92% of total metals produced in 2018 (Table 4.1), is both energy intensive and CO_2 emissions intensive. The sector is second to chemicals among all industries and shares about 20% of energy used in all industries around the globe. This is 8% of final energy used in all forms in 2019 and comprise of 16% (872 Mtce) of global coal demand (5530 Mtce) for making only coke (direct use for injection in blast furnace is additional to this) utilized almost entirely in this sector, 2.5% (90 bcm) natural gas and 5.5% (1230 TWh) of total electricity demand in 2019 globally in all sectors put together (IEA, 2019).

Iron is commercially available in five following different types.

1. Pure iron should have more than 99.8% iron and less than 0.008% carbon. It is ductile, magnetic and a conductor of heat and electricity. Pure iron is used as magnets, gaskets, fuse wire, welding rods and for magnetic resonance imaging (MRI), lighting conductors and making alloys of iron.
2. Wrought iron is close to commercial pure iron except the carbon percentage more than 0.008% but less than 0.1%. Total impurities less than 0.25%. Wrought iron can be heated, worked and reheated to get the required

TABLE 4.1

General Properties and Global Cardinal Data of Leading Metals

Metal	% of pure metal content in gross ore	% of concentrate content in gross ore (to convert 'metal content' towards 'gross ore run of mine')	% of concentrate content in 'ore concentrate' (to convert from 'ore concentrate' towards 'gross ore run-of-mine')	World Production (Ton in 2018)	% change in world production in 2018 with compared to 2014	% of total	Market Price (USD/kg)	Electric Conductivity (10.E6 Siemens/m)	Thermal Conductivity (W/m.K)	Density (kg/m3)	Melting or Deterioration Temperature (°C)	Market Value of Total Production (2020-21) (Million USD)	Share of Production in Method A (%)	Share of Production in Method B (%)	Embodied Energy of Metal Produced in Method A (GJ/Ton)	Embodied Energy of Metal Produced in Method B (GJ/Ton)	Total Embodied Energy of Metals Produced in 2018 (PJ/year)	Cradle-to-gate Contribution of 1 kg of Metal to Emissions of Greenhouse Gases (in CO_2-Equivalent) (kg CO_2/kg metal)	Energy Saving Potential by Use of Recycled Metal Instead of Its Primary Production from Ores (%)	Energy Saving Potential by Use of Recycled Metal Instead of its Primary Production From Ores (%)
Iron	43.3200	81.93000		1514832293	-3.65	0.918340	0.78	10.1	80	7900	1528	1181569	0.7	0.3	22.7	2.8375	25360.19	17	60-75	67.5
Aluminium	18.9800	67.55000		63237623	17.05	0.038337	2.39	36.9	237	2700	660	151264	1		211.5		13374.76	750	90-97	93.5
Silver	0.0340	2.55200		27698787	1.46	0.016792	985.91	62.1	420	10500	961	27308457			1500		41548.18	12000	96	96
Copper	1.0400	3.33000		20474372	10.31	0.012412	10.09	58.7	386	8900	1083	206505	0.8	0.2	33	64.5	804.6428	490	84-88	86
Zinc	8.3400	14.50000		12643000	-7.01	0.007665	2.95	16.6	116	7100	419	37240						10050	60-75	67.5
Lead	11.8600	16.52000		4640740	-13.39	0.002813	2.20	4.7	35	11300	327	10221						102	55-65	60
Gold	0.0002	0.06630		3367607	11.22	0.002042	66360.96	44.2	317	19400	1064	223477629	1		310000		1043958	4000000	98	98
Nickel	1.8300	23.45000		2256243	6.22	0.001368	17.33	36.9	237	2700	660	39092						1400	90	98
Tin	0.2400	0.33000		317862	3.27	0.000193	33.35	8.7	67	7300	232	10601						410		90
Uranium	0.0015	0.37440		65308	-3.72	0.000040												210		
TOTAL				1649533835		91.8340														

1 Oz= 0.02835 kg

1 Euro= 1.2186 USD

shape—and in the process, its grain structure, strength and durability improves. As it is malleable, until the advent of steel, wrought iron was the most sought-after iron to produce various items by forging, drawing, welding, rolling, extrusion and hammering.

3. Cast iron contains carbon at 2% or more and represent a vast variety of ferrous alloys depending on composition (silicon 1–3%, minor elements < 0.1% and other alloying elements < 0.1%), heat treatment and manufacturing process—mainly four: white, gray, ductile and malleable cast iron. Cast iron shrinks on solidification and so is extensively used in different casting processes to mass produce rails, wheels, axles, complicated shapes, etc.

4. Pig iron can contain 3.5–4.5% carbon, with sulfur less than 0.05%. It is an intermediate product to produce other forms of useful iron such as cast iron and steel. Pig iron is mostly produced in blast furnaces and mini-blast furnaces, and it is further processed in basic oxygen furnaces (BOFs), electric arc furnaces (EAFs), cupola furnaces, etc.

5. Direct reduced iron (DRI) is also known as sponge iron for its perforated look due to reduction of iron ore directly by either carbon or hydrogen (from natural gas or produced by electrolysis of water), thus justifying its name. DRI is also intermediate product, having no practical use other than to be processed further to produce steel. Because of vast variation of processing techniques on variety of ores, DRI could be of different physical and chemical properties as placed in Table 4.2.

4.2.1.2 Types of Metals: Aluminum

Aluminum is unique, being the most extensively available metal on Earth's crust (8%). It is infinitely recyclable (more than 75% of the aluminum ever produced from bauxite is still in use), features military-grade durability and corrosion resistance

---

TABLE 4.2
Typical Properties of Various Types of DRI

Elements	Unit	Gas based			Coal based	
		HBI	CDRI	HDRI	DRI Lumps	DRI Fines
Fe metallic	%	83–90	83–90	83–90	80–82	80–82
Fe total	%	89–94	89–94	89–94	90–92	90–92
Metallization	%	92–96	92–96	92–96	88–90	88–90
P	%	0.005–0.09	0.005–0.09	0.005–0.09	< 0.06	< 0.06
S	%	0.001–0.03	0.001–0.03	0.001–0.03	< 0.03	< 0.04
C	%	1.5–4.0	1.5–4.0	1.5–4.0	0.1–0.25	0.25–0.3
Product temperature	^{o}C	100	50	600–700	50	50
Typical size	Mm	30 × 50 × 110	4–20	4–20	4–20	0–4

Sources: www.engineeringtoolbox.com; www.ispatguru.com/iron-and-types-of-iron/#:~:text=Iron%20as%20an%20element%20is,(v)%20direct%20reduced%20iron

(a thin permanent coating of aluminum oxide protects the metal from any further corrosion or attrition), and is lightweight and energy efficient (as much less load in engines, automobiles, airplanes, etc.). Aluminum is now recognized as a sustainable metal of choice for modern transportation, building construction and packaging industries, etc.

Aluminum and its alloys, as in the case of iron, are of two types—wrought and cast categories, depending on production method. A product of aluminum, aluminum oxynitride (commercially named as ALON), produced in a chemical process, is the hardest known polycrystalline ceramic, recently developed as transparent sheet for many special uses including armor in defense, cockpits of aircraft, roofs of buildings (Figure 4.2), etc. Rolling, extruding, drawing, forging and a number of other more specialized processes belong to the wrought category. More than 200 aluminum and 400 aluminum-alloy types have been registered.

4.2.1.3 Types of Metals: Silver
Silver is the most electrical and thermal conductive metal. In 99.999% pure form, silver is used as silver paint or paste in printed circuit boards, electronics and solar photovoltaic junctions. Sterling silver (92.5% pure) is used as jewelry, silverware and other goods. Silver is weighed in troy ounces, one unit of which is equal to 31 gm or 0.031 kg; or in ounces, one unit of which is 28 gm or 0.028 kg.

4.2.1.4 Types of Metals: Copper
Copper has very high conductivity, close to that of silver, but the cost is almost a hundredth. Both cast and wrought copper are extensively used. As purity increases and porosity decreases, both conductivity and cost increases. Some impurities render copper with necessary hardness. However, performance is optimized by retaining certain impurity, as per required quality, at minimum cost. Because of porosity, cast copper with some impurity of iron can have International Annealed Copper Standard (IACS) levels at 93%, whereas wrought copper with reduced porosity and purity can have 102% IACS. Copper alloys with tin, zinc and nickel to make bronze, brass and cupronickel.

4.2.2 Common Applications of Metals

4.2.2.1 Common Applications of Iron and Steel: Cast Iron
Among the various properties of different types of cast iron, the most common is its hardness, which calls for careful selection of machine tool materials, such as coated carbides. Each type is suited for specific applications, depending on the specific properties as described in what follows.

4.2.2.1.1 Gray Iron Applications
Gray iron remains wear resistant even at limited lubrication; hence, it is used to produce cylinder heads, engine blocks, gear blanks, gas burners, housings and enclosures, etc.

4.2.2.1.2   White Iron Applications

White iron, being brittle and resistant to wear because of the chilling process of production, is used in the making of brake shoes, crushers, rolling mill rolls, pump housing, nozzles, etc.

4.2.2.1.3   Ductile Iron Applications

Depending on grades based on their properties, ductile iron is also wear resistant, along with its ability to be machined as well as high fatigue, yield strength and—of course—ductility. Some of its applications include heavy duty gears, crankshafts, hydraulic and automotive components, etc.

4.2.2.1.4   Malleable Iron Applications

Depending on grades based on their different microcrystalline structures, malleable iron is known for its ability to retain and store lubricants, its non-abrasive wear particles, and other abrasive debris in its porous surface. Some of its applications include chains and sprockets, heavy duty bearing surfaces, drive and axle components of trains, etc.

4.2.2.1.5   Compacted Graphite Iron Applications

Compacted graphite iron is used for its combination of the properties of gray iron and white iron that gives it high strength and high thermal conductivity; hence, it is suitable for diesel engine exhaust manifolds, blocks and frames, cylinder liners, train brake discs and high pressure pump gear plates.

4.2.2.2   Common Applications of Aluminum

Aluminum is known for its light weight (about one-third that of iron of same volume), high conductivity (more than half that of copper of same shape) and moderate cost (about three times that of iron of same mass). Unlike iron, aluminum does not rust. Thus, it is extensively used in electricity transmission with iron core to enhance the tensile strength, utensils, foils and cans for packing medicines and food, and an array of applications when alloyed with copper, magnesium, zinc, silicon and manganese. Duralumin, hindalium, Y-alloy, magnalium, etc., are used in aerospace, automobile, packaging and transportation industries for the properties have been enhanced with mechanical strength, fatigue resistance, etc.

4.2.2.3   Common Applications of Silver

Silver is extensively used in solar photovoltaic modules, electric vehicles, water filtration, ornaments and jewelry, high-value tableware and utensils in our day-to-day life, coins and bullion as an investment medium, special applications in electrical contacts and conductors, mirrors, window coatings, paint in printed circuit boards, as catalyst in chemical reactions, as an element or as a compound, such as in photographic and X-ray film, disinfectant, microbiocides for bandages and wound dressings, catheters, etc. [1]. In spite of recent innovations that decreased several fold the requirement of silver per solar photovoltaic (SPV) module, the phenomenal rise in

SPV power capacity deployment will continue to maintain the world demand of silver in the green energy sector at about 7,402 tons per year (CRU International, 2018).

4.2.2.4 Common Applications of Copper

While the price of silver is more than that of copper by about 100 times, the electrical and thermal conductivity of copper is marginally lower than that of silver and among the highest of all metals. Copper is thus the most favored metal when conductivity is important and the cost needs to be affordable, such as all wires and cables in motors, transformers, buildings and industry complexes; and also as lugs and contactors in switches, circuits, controls, panels, etc. Also, copper competes well with stainless steel (SS), for which conductivity matters more than strength of heat exchangers, heat pipes and such equipment.

4.3 ENERGY SYSTEMS IN METAL INDUSTRIES

Metals production from their ores involves the following steps and subsequent use of energy.

> Step 1: Exploration of ores from the mines—open cast (mostly) and underground mines—excluded from the scope of studies in this chapter, because the energy consumption in extraction of metals discussed here is much more than that for mining, as shown in Figure 4.3. Besides chemical energy of explosives, huge amounts of diesel may be consumed to operate the heavy machinery, huge excavation machines and haul trucks besides electricity, more for underground mining, may be from captive power plants to operate the conveyors. For lifting every kilogram of ore every kilometer, a theoretical minimum of 10 MJ energy, i.e. about 0.25 l of diesel or its equivalent coal, would be required—the practical quantities are much more and increase proportionally with the mass of ore and depth of mine.
>
> Step 2: Enrichment or beneficiation or concentration of ores, if required, as in case of copper, in particular. Depending on ore concentration, economy and available technology, the liberation size of ore is determined. Crushing (to about 5 mm size and bigger) followed by grinding (to about 0.1 mm and smaller) are increasingly energy-intensive processes. Electrical energy is also required to operate conveyors and sieves for screening and beneficiation, and then pumping the metal concentrates separated through flotation, a process within the scope of hydro-metallurgy.
>
> Step 3: Smelting of ore to produce metals still containing undesirable impurities—a high-temperature process of pyro-metallurgy. It can be exothermic, as in case of sulfide ores, aiding in cogeneration, or need carbon or hydrogen to remove oxygen, as in case of oxide ores; for example, in a blast furnace, iron (with carbon in excess of 2%) is produced by reducing iron ore hematite with coal dust and/or coke. Recycled metals, sometimes mixed with ores and/or their derivatives (such as DRI), are processed in EAFs. Thus, besides fossil fuels in pyro-metallurgy, large amounts of electricity also are used in conveyors, cranes, arc furnaces, compressors, fans, etc. Recent research

demonstrated use of hydrogen as a reducing agent instead of coal or coke, paving the opportunity to turn the metal industry green in near future, as both hydrogen and electricity may be produced in a renewable route.

Step 4: Purification of metals to lower the level of impurities within acceptable limits—for example, in BOFs from iron produced in blast furnaces in two phases—primary and secondary de-oxidation—the latter allowing the addition of some alloying elements, as per need. Oxygen used here at high pressure is produced in an energy-intensive process from atmospheric air using electricity. The chemical reaction is, however, exothermic to reduce the carbon percentage from more than 2% to the permission value.

Step 5: Alloying and/or heat treatment of the metals to attain desirable physical and chemical properties. Further adjustments after purification of metals by mixing with smaller amounts of alloying metals, as per requirements, are being done is ladle furnace using oil or gas as fuel.

Step 6: Manufacture various metal products we use, some of which are shown in Figure 4.1, which involves hot and cold rolling, galvanizing, trimming, electroplating, etc. The need for additional fuel for hot rolling can be drastically reduced by integrating hot rolling with continuous casting. Most of the processes here require electrically operated equipment and processes.

It is evident from Figure 4.3 that energy consumption in various stages of metal extraction are significantly different for different types of metals. In case of base metals, the mining and mineral processing consume almost same energy as that for smelting and refining, but for metals of more common use, such as iron, steel and aluminum, the later has much larger share with compared to former.

4.3.1 Generic Use of Energy

More or less, all metals are produced from its oxide ores by reduction. So, for carbon-intensive reduction processes, huge source of energy is consumed both as reducing agent and to provide the necessary heating effect. Probably the only exception is

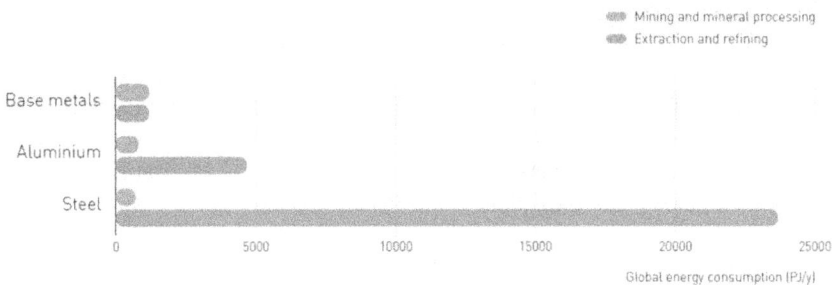

FIGURE 4.3 Share of energy consumption in mining and mineral processing, extraction and refining for metals.

TABLE 4.3

Direct and indirect energy inputs in generic steps of metal processing.

I	Indirect Energy Inputs				
	Production of flotation agents	Production of acids and solvents	Production of fabrication equipment	Production of jet fuels	Road infrastructure for collection trucks
IIa	Direct Energy Inputs and Their Corresponding				
	Electricity for blasting or crushing	Electricity and heat	Electricity	Combustion in jet energy	Heat for remelting
IIb	Generic Process of Metals Life Cycle from Extraction to Its Use until Recycling				
	Ore mining and processing	Smelting and refining	Manufacture and fabrication	Use in products	Recycling or disposal

sulfide ores, which are processed through an exothermic reaction and so a captive power plant is energized from this.

Typical share of energy consumption in different steps of metal processing is shown in Table 4.3.

4.3.2　Process Energy Systems

As shown in Table 4.1, the energy required to extract and utilize various metals from its ore to its productive use ("Embodied energy of metal") varies from 23 GJ/ton for iron to 310,000 GJ/ton for gold. This huge variation is because of the variation in concentration of metal in its ore (as high as 43% for iron to merely 0.00021% for gold) and the associated processes involving hydro-metallurgy and pyro-metallurgy. The relevant processes and associated energy use are explained in the following sections.

4.4　SMELTING AND REFINING

4.4.1　Ferrous Metal Industries

Ferrous metals primarily consist of iron and different varieties of steel and comprise of 91.8% of total metal production (Table 4.1). More than 8% of global final energy demand and 7% of energy sector CO_2 emissions (including process emissions) is emitted from ferrous metals industry (IEA, 2020). Demand for ferrous metals comes from construction and automobile sectors. This category includes: (i) iron and steel (integrated steel plants); (ii) mini steel plants; (iii) cast iron foundries; (iv) steel foundries; and (v) forging plants.

4.4.1.1　Iron and Steel: Integrated Steel Plant

The process of an iron and steel industry can be divided into various basic parts. First, raw material preparation consisting of sinter and pellet plants, iron making consisting of a blast furnace or an alternative route such as DRI, Corex or HIsarna units. Second, the steel making consists of OHFs, BOFs, EAFs and induction furnaces (IFs). Also, there are small-scale units based on DRI with scrap and only scrap recycle steel units using electric arc furnaces or IFs. Most of the steel is continuously cast for higher energy efficiency and productivity, except very few plants that still use ingot heating and rolling process. A schematic depicting the process is shown in Figure 4.4. Different stages of steel production are briefly described in what follows. The final shape is given in hot rolling mills, either reheating the continuously casted steel or directly charged hot as received from continuous casting line (CCL) for rolling process. Modern steel making also incorporates cold rolling at the finishing stage for grain orientation that improves the strength of steel, as hot rolling is still favored for higher productivity. A flow chart of different processes of steel making is shown in Figure 4.5, along with major energy flow.

4.4.1.1.1   Raw Material Preparation and Palletization

In the raw material preparation section, ore is washed, sized and fines of less than 200 mesh are agglomerated by sinter or pellet units in a process called palletization. These sinters or pellets of 9–16 mm size, after being dried in rotary kiln, are charged in a blast furnace. Larger-sized ore of 25–40 mm can be directly charged in to blast furnace. Whereas DRI, Midrex or HYL process produces the required output in the form pre-reduced iron, AKA sponge iron is charged in BOF/EAF/IF.

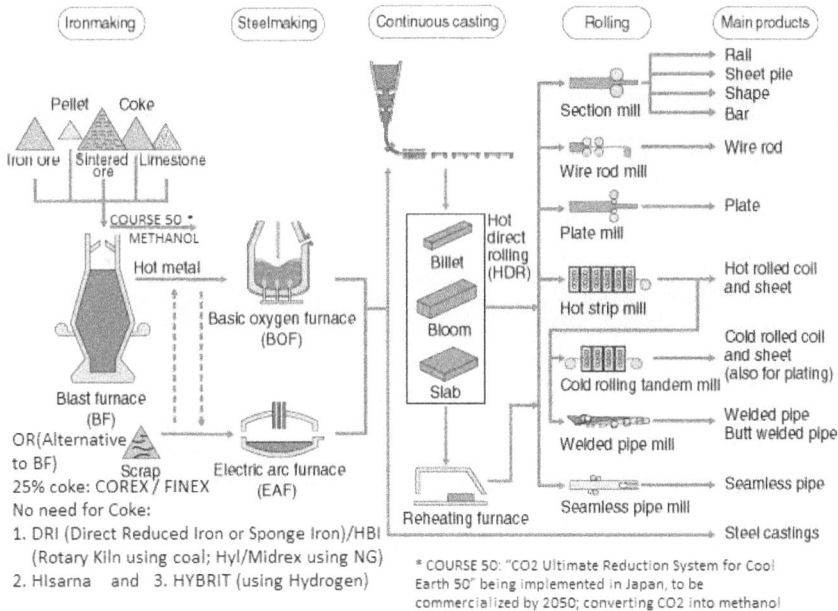

FIGURE 4.4   Steel production process flowchart.

Source: Adapted from "An Introduction to Iron & Steel Processing, Kawasaki Steel 21st Century Foundation"; www.jfe-21st-cf.or.jp/index2.html)

4.4.1.1.2   Iron and Steel Making

The raw materials prepared as per requirement and proportion is used in iron-making process in a blast furnace or Corex furnace. The output of the blast furnace system is hot metal. Output of DRI/MIDREX/HYL kilns are sponge iron or DRI can be used in AEFs and IFs, along with or without hot metal.

In the steel making process, a basic oxygen furnace (BOF), besides hot metal, utilizes scrap in certain proportion and blown with high pressure 99.9+% pure oxygen to reduce the carbon content from hot metal which possess in excess of 4% carbon. EAFs and IFs utilize electricity to melt and required components are added to form the desired composition of steel, then the steel is shaped in the casting process to produce various steel products. The rest follows the same process.

The Circored process is to produce hot briquetted iron (HBI) from iron ore fines being reduced with hydrogen generally obtained in specially designed reactor reformed using natural gas as fuel (Steel online Open Energy Monitor, 2021). Alternately, biomass can be used to produce the required gas. HBI has about 95% metallization and can be used along with scrap to make iron in EAF. Thus total energy required will be 14.7 GJ/ton for electricity and 6.5 GJ/ton for biomass.

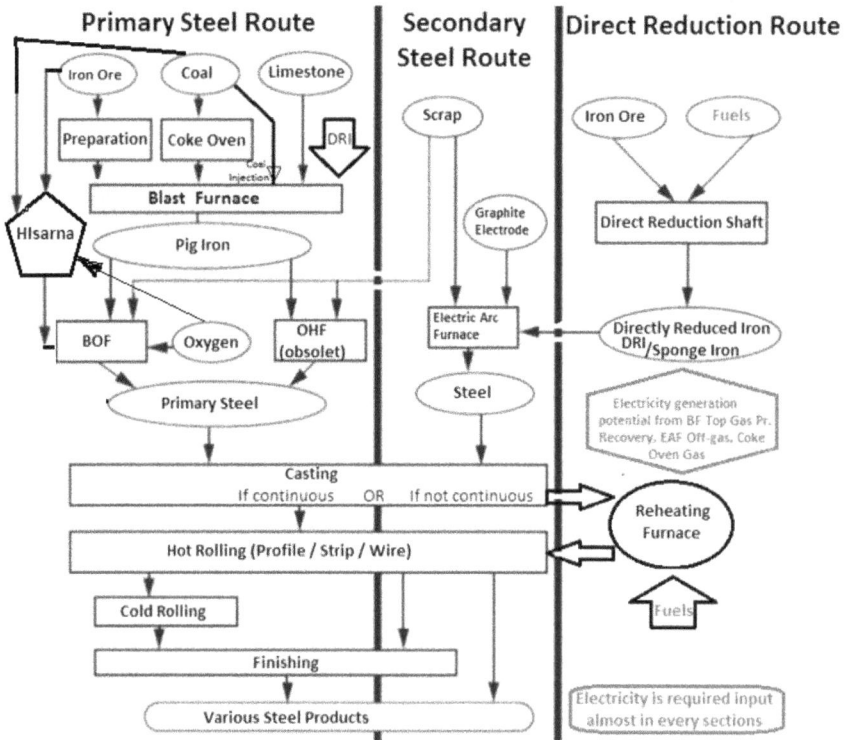

FIGURE 4.5   Flow chart of different process routes to steel making and energy flow in it.

Source: Adapted from BEE (2015)

HIsarna is combination of three processes—heated screw coal pyrolysis feeder, cyclone converter furnace (CCF) and HIsmelt vessel—to achieve 30% higher energy efficiency, lower CO_2 emissions at 100% pure form, almost half the cost of blast furnace and yet producing iron directly from pulverized ore, coal, etc., eliminating the need for sinter, coke, etc. (Figure 4.6). This alternative to the BF route of producing primary steel was developed during 2012–2018 at IJmuidin plant of Tata Steel Europe under the ULCOS (ultra low–carbon dioxide steelmaking) project in Europe and is expected to be commercialized within a couple of years.

Another innovative initiative started in 2016 in Luleå, a small town in northern Sweden, which aims to produce DRI by 2027 using hydrogen instead of carbon and thereby turn the steel industry toward a zero-emission industry under HYBRIT project launched as a joint venture between the utility Vattenfall, iron ore producer LKAB and steel maker SSAB. What is more interesting is to enhance the renewable energy power generation and distribution capacity required if this initiative to completely delink the steel industry from carbon emission. For example, in Sweden, this would call for additional electrical load of 10 GW to be met by installation of new renewable power plant and electricity distribution system, with existing total capacity of 40 GW in the national grid. Presently, three pilot projects are aiming

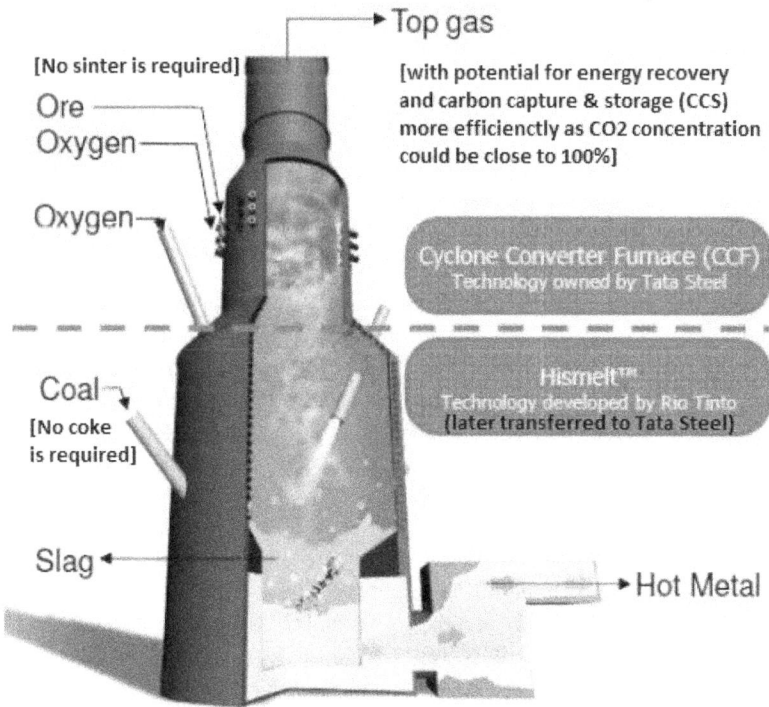

FIGURE 4.6 HIsarna technology for efficient and cleaner production of steel.

Source: PMI (2018)

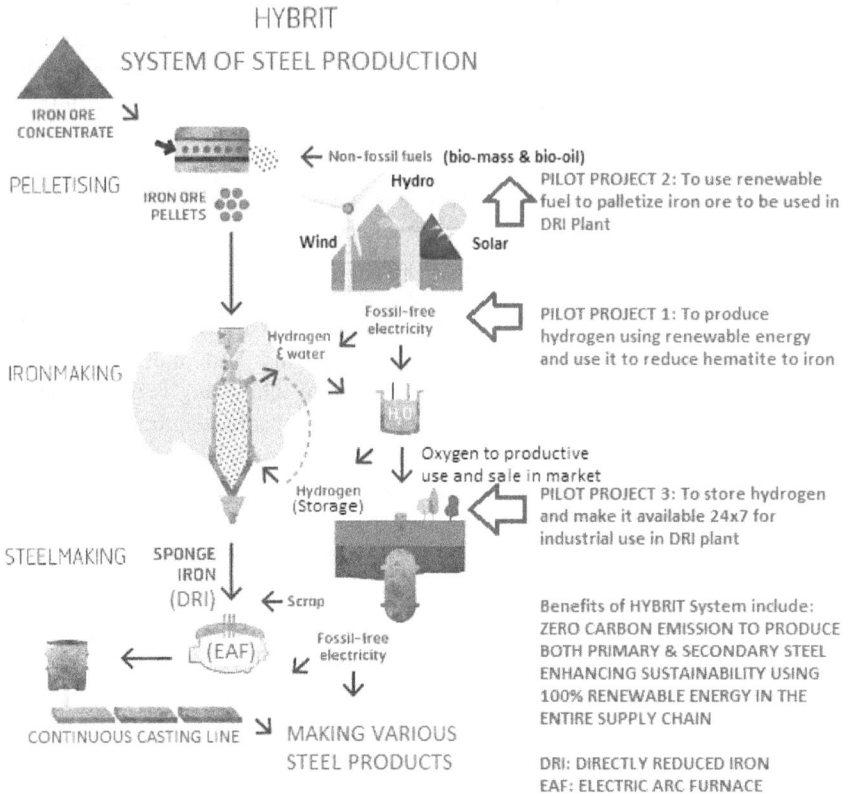

HYBRIT
SYSTEM OF STEEL PRODUCTION

IRON ORE
CONCENTRATE

PELLETISING

IRON ORE
PELLETS

← Non-fossil fuels (bio-mass & bio-oil)

Hydro

Wind   Solar

PILOT PROJECT 2: To use renewable
fuel to palletize iron ore to be used in
DRI Plant

IRONMAKING

Hydrogen
& water

Fossil-free
electricity

PILOT PROJECT 1: To produce
hydrogen using renewable energy
and use it to reduce hematite to iron

Oxygen to productive
use and sale in market

Hydrogen
(Storage)

PILOT PROJECT 3: To store hydrogen
and make it available 24x7 for
industrial use in DRI plant

STEELMAKING   SPONGE
IRON
(DRI)   ← Scrap

(EAF)   Fossil-free
electricity

Benefits of HYBRIT System include:
ZERO CARBON EMISSION TO PRODUCE
BOTH PRIMARY & SECONDARY STEEL
ENHANCING SUSTAINABILITY USING
100% RENEWABLE ENERGY IN THE
ENTIRE SUPPLY CHAIN

CONTINUOUS CASTING LINE   MAKING VARIOUS
STEEL PRODUCTS

DRI: DIRECTLY REDUCED IRON
EAF: ELECTRIC ARC FURNACE

FIGURE 4.7 Schematic of HYBRIT project being executed by Vattenfall, LKAB and SSAB of Sweden [2].

to produce output complementing each other: (i) to produce fossil-free electricity to electrolyze water into hydrogen and oxygen, use hydrogen to produce DRI from hematite; (ii) 100% renewable fuel operated palletization plant; and (iii) Hydrogen storage facility to make it available during industrial production of steel from mining to finished product delivery using only 100% renewable fuels, as shown in the schematic diagram in Figure 4.7 (Energypost, 2020).

4.4.1.1.3   Hot Strip Mill (HSM)

Metals are extensively used in sheets which are also galvanized or colored. The process involves hot rolling (for faster production) and cold rolling (for quality and precision) after casting of metals in ingots, billets or blooms. Considering the energy efficiency and productivity, research in the last century has established the acceptance of several technologies that integrate the casting, heating (if required), rolling, striping, galvanizing/coloring and coiling operations to minimize wastage and also maintain the required flexibility in order to meet the needs of the market. CSP, its derivative CSP Nexus and a similar technique, in-line strip production (ISP), and Finex process are prime examples. Technology pioneers such as Arvedi, SMS

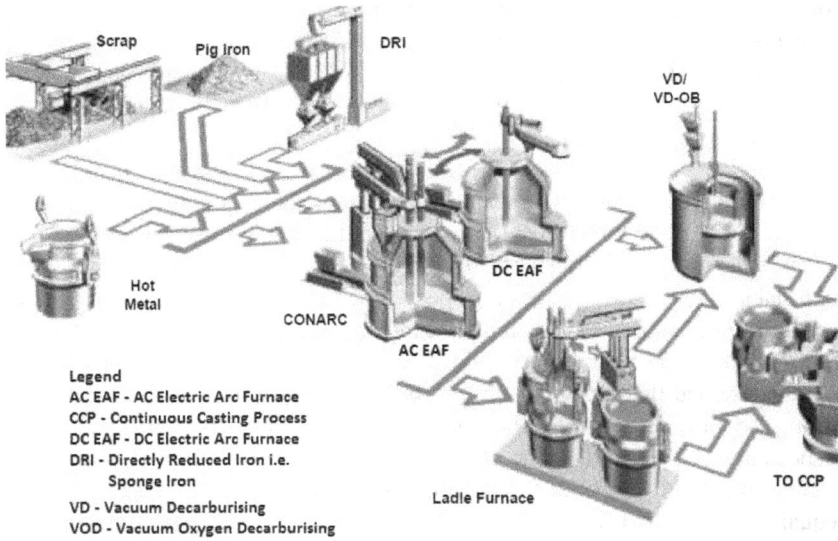

FIGURE 4.8   A typical hot strip mill (HSM) process flow diagram.

Source: Schmale, Kersten, and CONARC® (2003)

Group, Mannesmann Demag and VAI paved the way for such productive coherence of manufacturing processes with minimal stages for zero waste objectivity. CSP is a continuous process that significantly reduces the production workflow from liquid phase metals to the finished hot rolled strip. ISP produces hot rolled coil down to finished gauges of 1 mm or lower. CSP Nexus was expected to be all set to go for production in 2021 for the first time in the world by Steel Dynamics Inc. It is a modular system with maximum flexibility for most delicate tailor-made solution and yet can be expanded with, for example, a second casting strand or lateral slab feeding or induction heating and even high-speed shear for endless operation, all with the option for fully electricity run to make it even 100% green production line.

Typical features of an HSM plant (Figure 4.8) are as follows.

- Capacity: 3.0 MTPA.
- Twin shell CONARC furnace capable of using different combinations of charge mix of hot metal, sponge iron and scrap.
- Ability to roll as thin as 1.2 mm with 1,250 mm width.
- Capable of producing ultra-low-carbon steels/Electrical steels through VOD/VCD/VD routes.
- HR as a substitute for CR in certain applications due to superior surface quality and thinner gauges.
- Rolling mill has HGC, AGC, CVC and roll bending in all six stands.
- Roll shifting facility in all stands of +/− 100 mm for better profile.

Following is a brief description of the process.

In HSM, HR coil is produced from the liquid metal. The main raw materials for this section are hot metal from blast furnace, sponge iron from SIP and electrical power. Basically, it is a continuous casting process. Continuous casting is the process whereby molten steel is solidified into a "semifinished" billet, bloom or slab for subsequent rolling in the finishing mills.

The continuous casting machine receives steel from electric arc furnace in hot liquid form via a ladle feeding a turndish. Solidification begins in the copper mold portion of CCM, which release red-hot strand for finished rolling.

Next in the process is a tunnel furnace. The roller hearth furnace receives cut-to-length slabs from the caster at casting speed. The furnace operational speed and the process heating control is speed-matched to the CSP caster and thin slab temperature, as monitored at the entrance to the furnace. Length of the soaking furnace plant is so dimensioned that a roll change can be done also for the second stage during casting. The buffer capacity for the first stage is 20–35 minutes, depending on casting speed and slab advance at the beginning of a possible interruption in mill area. Under normal steady operational conditions the roller hearth furnace is required only to equalize the thin slab's temperature profile, since its inlet temperature of approximately 1100°C (only at surface) is what is required for discharge into the mill. At the discharge end of the furnace, thin slabs have a temperature differential of ± 10°C only along the slab's length, thickness and width. The excellent temperature profile of the thin slab entering the CSP mill provides optimum conditions for the subsequent rolling process, and thus ensures an excellent profile and flatness of the hot rolled product.

Electrical power, propane and LDO are used in different stages of the production process.

4.4.1.1.4 Rolling Mills and Finishing

In rolling mills, both hot rolling and cold rolling are used, as per requirement. In the case of hot rolling, steel is heated and passed through heavy rollers to reduce the thickness or provide the desired shape. Hot rolled steels are used primarily for grain orientation and also reduce thickness and smooth surfaces faster than cold rolling. Cold rolling mills are known for better surface finishes and packed grain orientation, which make the product more durable, but consumes more electrical energy and thermal energy is used for pickling, followed by treatment in a tandem mill. The final process involves finishing of the product, whereby annealing and surface treatment are done particularly for the products from hot rolling mills. This principle applies to all metals—both ferrous and non-ferrous.

4.4.1.2 Mini Steel Plant

A mini steel plant is a small plant for the manufacture of steel. These plants use pig iron or scrap iron as their raw materials. Mini steel plants in India are completely dependent on electric power and hence do not cause pollution.

Here and elsewhere, a number of both ferrous and non-ferrous metal manufacturing processes are conducted which are highly energy intensive and suitable for EE&C. These sections include cast iron foundry, steel foundry, project and design, molding and core making, melting and casting, de-molding, shot-blasting, sleeves-feeders cutting, heat treatment, deburring, quality control and inspection, machining, forging, etc. Energy consumption patterns and conservation are discussed for some of these processes in the following section.

4.4.2 NON-FERROUS METAL INDUSTRIES

Non-ferrous metals primarily consist of aluminum, silver, copper, zinc, lead, gold, nickel, tin and uranium, comprising the rest, i.e. only 8.2% of total metal production in the world. Out of these, only aluminum, silver and copper have shares of more than 1% (i.e. 3.8%, 1.7% and 1.2% respectively) and are also discussed in the following subsections.

4.4.2.1 Aluminum Manufacturing Process Energy System

Aluminum is the most abundant among all metal elements on Earth's crust but is extremely reactive, mostly found as bauxite—an oxide of aluminum hydrated and mixed with other impurities, such as compounds of other elements. Thus, extraction of aluminum involves first, the purification of bauxite in to alumina (pure oxide) and then its electrolysis to obtain aluminum metal in liquid form which can be given the shape of billet, bloom or bars for productive used latter. Approximately 6 tons of bauxite produces 2 tons of alumina which further can produce 1 ton of aluminum metal. A more energy-efficient method would be to go for continuous casting and subsequent metal forming processes, thereby avoiding or minimizing the use of energy for reheating of billet or bloom or bar. Aluminum is the second-most used metal in industries and day-to-day life for its excellent properties and versatilities (Table 4.1). Production of alumina (metallurgical) and primary aluminum are shown in Table 4.4. The ratio of alumina to aluminum production has been found to vary from region to region, but as expected, 1.96 (very close to 2) globally as we know that about 2 tons of alumina is required to produce 1 ton of aluminum. The regional variation indicates internal trade, as the electrolysis of alumina is extremely energy-intensive process and is economically viable at places where adequate electricity is available at cheaper rates, for example, near hydropower plants. In India, with one of the lowest per capita use of this metal in world, the use of aluminum is found in decreasing order in the following sectors: electrical, transport, construction, consumer durables, machinery and equipment, and packaging. Aluminum is endlessly recyclable, and there is no loss of properties or quality during the recycling process. More than 50% of aluminum produced during modern civilization is still in use.

Purification of bauxite to alumina is popularly done via the Bayer process. Extraction of 99% pure aluminum is achieved through electrolysis of alumina in a specially designed process named after its inventers, Hall–Héroult. To meet specific demand of extremely pure aluminum (99.99%), the Hoope process is put to work.

Bauxite contains 30–54% alumina (Al_2O_3), along with a mixture of silica, various iron oxides and titanium oxide as impurities which need to be removed in the Bayer process (Figure 4.9), whereby it is digested in a hot solution of caustic soda (NaOH at 175°C) so that only alumina is converted to its hydroxide and dissolves in water according to the following equation.

TABLE 4.4

Global Alumina (Metallurgical) and Aluminum Production during 1 July 2020–30 June 2021

S N	World Regions	Annual Production of Aluminum		Annual Production of Alumina		Alumina/Aluminum
		(Thousand Metric Tons of Aluminum)	Share (%)	(Thousand Metric Tons of Aluminum)	Share (%)	
1	Africa	1,582	2%			
1A	Africa and Asia (except China)	5,882		12,410	10%	2.11
2	North America	3,941	6%	2,084	2%	0.53
3	South America	1,079	2%	11,459	9%	10.62
4	Asia (except China)	4,300	6%			
5	Western Europe	3,340	5%	4,079	3%	1.22
6	Eastern and Central Europe	4,136	6%	4,620	4%	1.12
7	Oceania	1,876	3%	20,819	16%	11.10
8	GCC	5,796	9%			
9	China (estimated)	38,658	58%	70,612	54%	1.83
10	ROW estimated unreported	2,000	3%	4,412	3%	2.21
	Total	66,708	100%	130,495	100%	1.96

Notes: ROW = Rest of World; GCC = Gulf Cooperation Council

Source: Website of the International Aluminium Institute (www.world-aluminium.org/statistics/, accessed on 12 August 2021)

$$Al_2O_3 + 2\ OH^- + 3\ H_2O \rightarrow 2 \times [Al(OH)_4]^-$$

Pure aluminum hydroxide is precipitated as a white fluffy solid as the solution is cooled down to ambient conditions and is separated and calcined at 980°C in a rotary kiln to produce pure alumina. However, in the process, a large amount of impurities—called red mud—is separated out and must be stored in a holding pond, as it is highly basic and so far remains a challenge for productive use. As its color signifies, "red mud" is rich in oxides of iron and also a bit of titanium.

The Hall–Héroult process (Figure 4.9) is the major and most popular industrial process for the production of aluminum of required purity from alumina obtained by the Bayer process. Aluminum electrolysis with the Hall–Héroult process consumes a lot of electricity, but alternative processes were always found to be less viable economically and/or ecologically. It involves dissolving alumina in molten cryolite, and electrolyzing the molten salt bath to obtain pure aluminum metal. The melting point of alumina is above 2000°C, whereas that of cryolite (Na_3AlF_6, sodium hexafluoro-aluminate, an uncommon mineral) is about 1000°C. Some aluminum fluoride (AlF_3)

is also added as a small quantity of alumina is dissolved in cryolite solution contained in a cathode holder to further reduce the temperature of the electrolysis bath for higher energy efficiency and operating life. The molten (without water) solution is contained in cathode vessels, where more than 99% pure liquid aluminum is precipitated and collected by siphon pipes to avoid delicacy of pumping liquid aluminum which may thus be solidified as billet or more energy-efficiently utilized through continuous casting process, similar to what has been discussed in case of iron and steel. Oxygen is accumulated in the sacrificing carbon anode and is liberated as carbon dioxide, which can be utilized after removing the contaminants such as fluorine, which is highly reactive. As the required voltage difference is within 5 V, a large number of such (Hall–Héroult) cells are connected in series to receive high-ampere DC supply from the rectifier system.

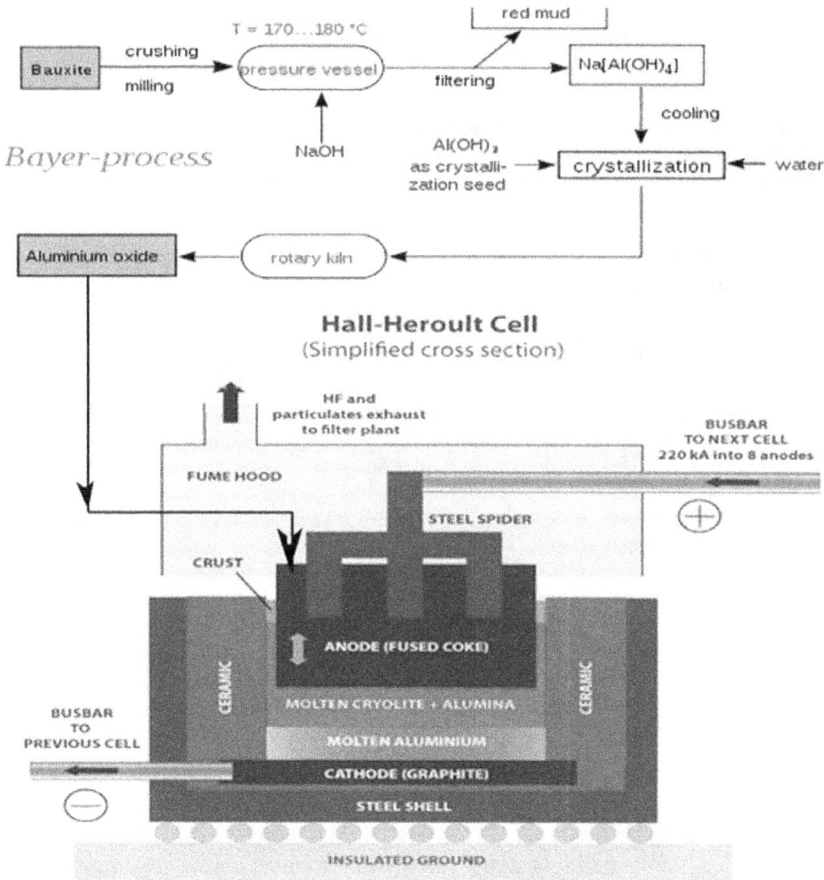

FIGURE 4.9   Energy and materials flow: manufacturing aluminum from bauxite.

4.4.2.2 Silver Manufacturing Process Energy System

Silver is produced along with other metals such as zinc, copper, lead and antimony. For example, a typical silver ore might contain only 0.085% silver, 0.5% lead, 0.5% copper and 0.3% antimony. After a concentration process, called flotation separation, the concentrate may be enhanced to contain 1.7% silver, 10–15% lead, 10–15% copper and 6% antimony. Interestingly, approximately only 25% of the silver produced globally comes from ores actually mined for their silver value; the other about 75% comes from ores that have as their major metal value other than silver, such as either lead, copper or zinc. Mostly, these ore minerals are sulfides; typically, lead is present as galena (PbS), zinc as sphalerite (ZnS) and copper as chalcopyrite ($CuFeS_2$). In addition, the mineralization usually includes large amounts of pyrite (FeS_2) and arsenopyrite (FeAsS). As a reference, therefore, silver manufacturing will be discussed along with that of copper in the following section.

4.4.2.3 Copper Manufacturing Process Energy System

Manufacturing of copper from its ore has some similarity with that of aluminum from bauxite, except for some differences, most notably the following.

1. Unlike oxide of aluminum, copper is obtained mostly from its sulfides.
2. Concentration of copper in its ore can be significantly lower, even as low as 1%.
3. Therefore, an additional step for its beneficiation is essential to increase concentration to at least 25% for consideration in smelting and refining.
4. As sulfur of ore combines with oxygen in a exothermic reaction, copper smelting is somewhat lower in energy intensity with compared to aluminum, though the need for two stage oxidation somewhat lowers the difference.
5. The byproduct, sulfur oxide, can be utilized to produce sulfuric acid.
6. The refined copper after electrolysis is in solid form, unlike aluminum, which is produced in liquid form and so can be taken directly to a continuous casting route; whereas, solid copper, so produced, needs to be melted first before casting.

The energy interactions in various stages of copper manufacturing are as follows.

1. Concentrating: cone crushers are used to first break the ore lumps into smaller pieces, which are further ground in to finer pieces (3 mm) in a series of rod mills and finally mixed with water to make a slurry of fine particles (0.25 mm). This slurry is mixed with frothers, such as pine or organic oil, to separate copper particles from huge amounts of dirt and other impurities by flotation. Electricity is used to run the motors for mills and pumps. Froths are collected in a pond, settled and dried as output. Normally, gold and silver also are found in smaller amount in the output of copper with concentration more than 25%.

2. Smelting: this process traditionally involves two stages of operation in two furnaces. Both the operations utilize a single furnace in some modern plants, which combines the two processes. In the first stage, the sulfur in the concentrated output is burned with oil and oxygen-enriched air to increase the concentration of copper to about 60% (called matte). The byproduct, as mentioned earlier, will be sulfuric acid. Iron present is converted to oxide and removed as slag. In the second stage, in reactor furnace, the process is repeated, with pure oxygen, to increase the concentration to more than 99%, called "blister copper." It is now ready for further purification to 99.9% copper as conductor grade by the electrolysis method.

3. Refining: conductor-grade copper is produced from copper blister in two stages: thermal and electrical refining.

   i. Thermal: in thermal refining process, following the process similar to the proceeding reactor furnace, copper is purified to 99.5% and poured in the anode mold which, in the remaining 0.5%, also contains gold, silver, selenium and tellurium.

   ii. Electrolysis: molded copper anodes are used as sacrificing anode dissolved in a solution of $CuSO_4$ and H_2SO_4, and pure copper is deposited on the thin copper sheet used as cathode in the electrolytic cells supplied with high-ampere DC. The required temperature is maintained by adjusting the depths of anode that alter the resistance and its heating effect. To attain the productivity, it is a practice to join more than 1,000 cells at a time. Copper obtained at cathode is 99.95–99.99% pure.

   iii. Byproduct from slime: the elements, other than copper, which were present in the sacrificing anode mold get deposited at the bottom of the cell tank as slime which is collected from time to time and further processed to extract gold, silver, selenium and tellurium.

4.5   ENERGY AND MASS BALANCE

Energy and mass balance of metal manufacturing processes have the following benefits.

1. Mass and energy balance follow the law of conservation of mass and the law of conservation of energy, which can be applied to any and every system of all the metal manufacturing processes and declares that energy and mass cannot be created or destroyed, but only can take part in processes to change its form/composition. Thus, it is simple to maintain though calls for various measuring gadgets and facilities. With the advent of energy information system (MIS) and the IoT, mass and energy balance could be easier.

2. It helps to identify the point of losses of two important resources: energy and materials.

3. A check and control mechanism could be in place to monitor, assess and optimize the process by controlling the mass and energy flow.

4. Energy and mass interaction data can be visually displayed in the modern distributed control system (DCS) and also even in the flow charts of various processes maintained at the floor level.
5. Typical and standard values of energy and mass data of various metal manufacturing processes are available in the Technical Manual of the same and the Standard Operating Practices (SOPs).

Huge potential exists for production of metals with lower consumption of energy as the ratio of actual energy consumption to that of theoretical energy consumption in general is high for most metals. It is varying from more than 2 for iron to more than 25 in case of nickel, as shown in Figure 4.10.

The metals extraction processes around the world are run at varying efficiency levels, as shown in Figure 4.11, which shows the life cycle energy requirement in MJ/kg of produced metals. While the aluminum industries are operating more toward highest efficiency, copper and steel have more scope, on average for improvement by about 50% of the present values. The range and average of life cycle energy required for primary production of selected metals (in MJ/kg of produced metals) are shown in Figure 4.11. The average values in most of the cases are skewed toward lower value of the range, indicating an increasing practice of energy efficiency in the metals sector as energy cost has become the more important controllable criteria for competitive advantage.

Ratio of Actual/Theoretical Energy Consumption

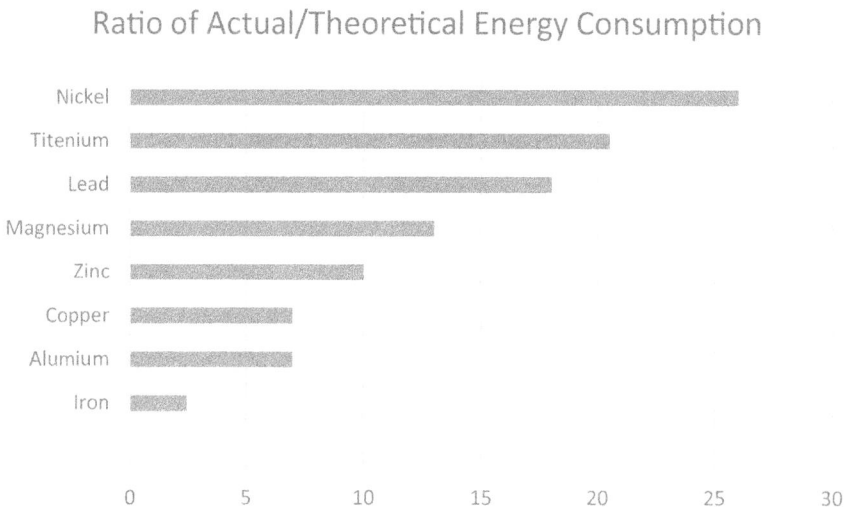

FIGURE 4.10   Ratio of actual to theoretical energy consumption in metal production.

Source: "Environmental Risks and Challenges of Anthropogenic Metals Flows and Cycles"; Report #3 of the Global Metal Flows Working Group of the International Resource Panel of UNEP; 2013

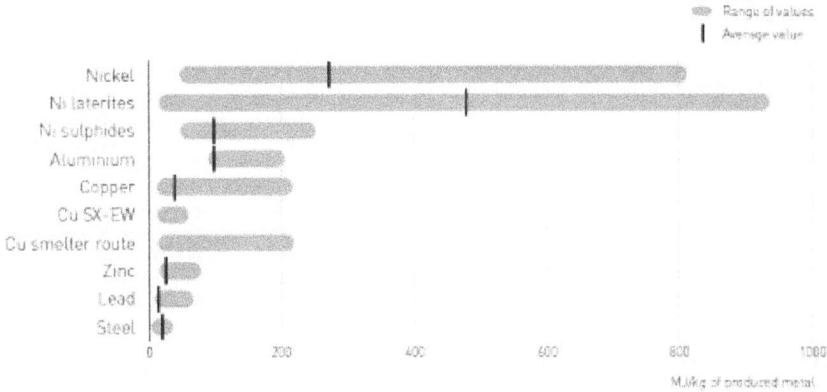

FIGURE 4.11 Life cycle energy required for primary production (in MJ/kg of produced metals), selected metals—range and average.

Source: UNDP (2013)

4.5.1 DIGITIZATION OF THE STEEL MAKING PROCESS AND INCORPORATION OF THE CIRCULAR ECONOMY FOR ENERGY EFFICIENCY

As most of the emissions of metals industry are from the ferrous processes, several techniques are in progress to drastically reduce such emissions and make the production line much greener. For example, the Finex process is a lower-cost, more environmentally friendly alternative to traditional blast furnaces for producing hot metal. Widespread participation in programs to reduce CO_2 emissions such as ULCOS in Europe, Course 50 in Japan and the AISI CO_2 Breakthrough Program in North America, are among the noteworthy efforts. HIsarna and HYBRIT processes have been mentioned in this chapter. Successful trial has been made to produce 100% green resourced steel. Unlike the conventional "linear economy" model, the circular economy, also known as the productive economy, employs a system whereby everything is treated as resource for at least one productive economy activity in the metal industries. Neither any material nor any form of energy can be treated as waste, even in the entire life cycle of the metal products. A typical energy and material flow diagram and its values is shown in Figure 4.12.

4.5.2 ENERGY REQUIREMENTS IN ALUMINUM PRODUCTION

Energy cost can be 30–70% of the aluminum manufacturing cost, depending on the price of materials, human resources and energy. Therefore, more urgency needs to be given to reduce its energy intensity, particularly in the smelting process. The values for different regions of the world for more than a decade have not significantly changed, except for China and Oceania (i.e. Australia). Probably because of COVID-19, the decreasing trend also tended to remain the same or slightly increased in case of all the regions of the world in 2019 and 2020. The average figure for the

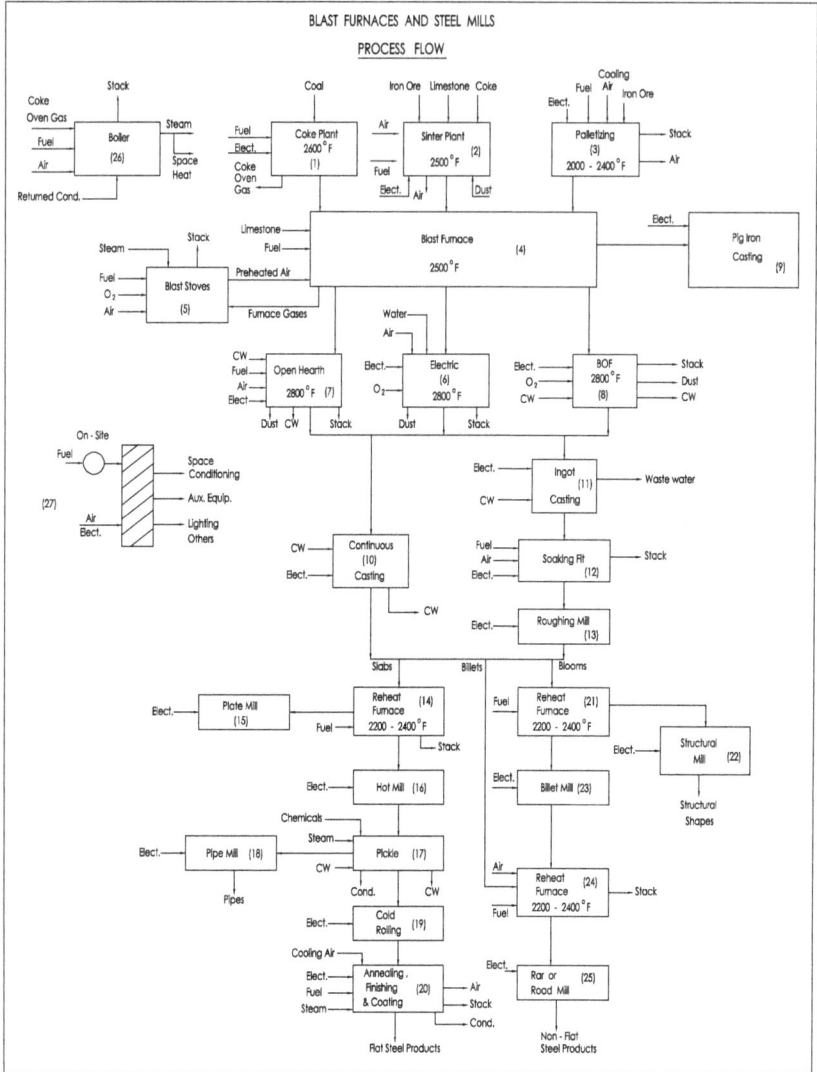

FIGURE 4.12 Typical energy and mass flow of blast furnace and steel mills in an iron manufacturing plant.

Source: Brown, Harry L. et al. (1985) Energy Analysis Of 108 Industrial Processes, Drexel University Philadelphia. PA 19104, US Department of Energy.

world is between 14,000 kWh/ton and 14,500 kWh/ton, whereas for China, it is close to 13,500 kWh/ton. In view of the lower theoretical value of 3,216 kWh/ton, the efficiency varies from a low of 38.8% (South America, 2011) to a high of 46% (China 2019).

TABLE 4.5A

Typical Energy and Mass Flow of Blast Furnaces and Steel Mills in Iron Manufacturing Plants

Unit operation			Inlet				Outlet			
No	Description	Temperature (°f)	Flow	Temperature (°f)	Mass (lb)	Energy (BTU)	Flow	Temperature (°f)	Mass (lb)	Energy (btu)
1	Coke ovens	2600	Coal	75	0.747	0	Loss			195
			OFG		0.106	124.5	Coke	150	0.528	10
			COG		0.033	622.5	Breeze	85	0.028	0
			Tar		0.005	83	Cog	85	0.119	1.2
			Fuel			58.1	LT Oils	85	0.007	0
			Quenching Water	75	1.43	0	Tar	85	0.028	0.1
			Gas Water	75	2.6	0	Con D	180	0.136	14.3
			Electric			20	NH3 Water	120	0.037	1.7
			Air	75	0.710 0.0	Granting Water	Gas water	200	2.6	330
			Exotherm			122	Quenching water	200	1.21	151.3
			Steam	350	0.136	150	Quenching vapour	212	0.22	218.4
							Stack	700	0.86	255
2	Sinter Plant	2500	Ore	75	0.296	0	Sinter	500	0.385	54
			Dust	75	0.036	0	Dust	500	0.075	4.9
			Limestone	75	0.116	0	Cool air	250	3	126
			Scale	75	0.23	0	DDST air	95	2.2	10.6
			Coke		0.023	293.4	Stack	400	0.92	78.1
			Fuel			86.7	Loss			35.5
			Electric			20				
			Endotherm			-91				
			Air	75	6.12	0				

(*Continued*)

TABLE 4.5A Continued

Unit operation

			Inlet				Outlet			
No	Description	Temperature (°f)	Flow	Temperature (°f)	Mass (lb)	Energy (BTU)	Flow	Temperature (°f)	Mass (lb)	Energy (btu)
3	Pelletizing	2400	Ore	75	0.665	0	Pellets	500	0.661	92.7
			Coke		0.012	145.6	DEDST Air	95	1.8	8.6
			Fuel			76.7	Stack	500	1.5	177
			Exotherm			96	Loss			60
			Electric			20				
			Air	75	3.3	0				
4	Blast Furnace	2600	Coke	75	0.521	0	Big Iron	2550	0.838	429.7
			Scrap	75	0.082	0	Slag	2600	0.41	317.7
			Ore	75	0.312	0	Dust	420	0.026	2.5
			Sinter	500	0.395	54	BFG	450	3.142	390.1
			Pellets	500	0.661	92.7	CW	85	35.83	358.3
			Limestone	75	0.258	0	Stack	600	0.175	30
			Air	1600	2.2	805.2	Loss			281.1
			CW	75	35.83	0				
			COG		0.001	20.7				
			Tar		0.002	29.1				
			Exotherm (coke)			3163				
			Endotherm			-2690.5				
			Electric			60				
			Fuel			275.2				

Source: Brown. Harry L et al. (1985) Energy Analysis Of 108 Industrial Processes., Drexel University Philadelphia. PA 19104, US Department of Energy

TABLE 4.5B

Typical Energy and Mass Flow of Blast Furnaces and Steel Mills in Iron Manufacturing Plants

No	Description	Temperature (°f)	Flow	Temperature (°f)	Mass (lb)	Energy (btu)	Flow	Temperature (°f)	Mass (lb)	Energy (btu)
5	Blast stoves	1600	PRH air	75	2.2	0	Hot air	1600	2.2	805.2
			Steam	800	0.32	436.6	Stack	300	1.65	96.5
			BFG (Fuel)		0.56	660	Cond	100	0.32	8
			CW	75	13.3	0	Loss			54.3
			Air	75	1.09	0				
6	Electric furnace	2800	Scrap	75	0.26	0	ML T steel	2800	0.24	99.5
			Air	75		0	Slag	2800	0.02	18.2
			O2	75	0.01	0	Off Gas	500	0.546	99.1
			Electric			255	CW	95	2.885	57.7
			Carbon		0.002	24	Water	200	0.09	11.2
			CW	75	2.885	0	Loss			35.3
			Exotherm			42				
			Qnch wtr	75	0.136	0				
7	Open heart	2800	Pig iron	2550	0.2	102.6	ML T Steel	2300	0.3	124.4
			Scrap	75	0.11	0	Slag	3000	0.05	46.9
			Ore	25	0.02	0	Steam	800	0.15	204.8
			Limestone	75	0.02	0	Dust	500	0.01	1.1
			Air	75	1.1	0	CW	95	4.8	86

(*Continued*)

TABLE 4.5B Continued

No	Description	Temperature (°f)	Flow	Temperature (°f)	Mass (lb)	Energy (btu)	Flow	Temperature (°f)	Mass (lb)	Energy (btu)
			CW	75	4.8	0	Stack	500	1.2	229.8
			Con D	100	0.15	3.8	Loss			75
			Fuel			550				
			Electric			20				
			Exotherm			92				
8	Basic oxygen furnace	2800	Pig Iron	2550	0.61	312.7	MLT steel	2300	0.745	306.9
			Scrap	75	0.27	0	Slag	2800	0.125	117.3
			Limestone	75	0.04	0	Dust	500	0.01	1
			O2	75	0.05	0	CW	95	9.3	186.5
			Air	75	0.42	0	Off gas	600	0.67	257
			CW	75	9.3	0	Loss			95
			Qnch Wtr	75	0.16	0				
			Electric			120				
			Exotherm			632				
9	Pig iron casting	75	Pig iron	2550	0.02	6	Cast iron	75	0.02	0
			Electric			30	Loss			36

Source: Brown, Harry L et al. (1985) Energy Analysis Of 108 Industrial Processes, Drexel University Philadelphia, PA 19104, US Department of Energy

TABLE 4.5C

Typical Energy and Mass Flow of Blast Furnaces and Steel Mills in Iron Manufacturing Plants

No	Description	Temperature (°F)	Flow	Temperature (°F)	Mass (LB)	Energy (BTU)	Flow	Temperature (°F)	Mass (LB)	Energy (BTU)
10	Continuous casting	1600	ML T steel	2800)	0.08	26	CW out	90	0.8	9.2
			CW In	75	0.8	0	Ingot	1600	0.08	15
			Electric			30	Loss			31.8
11	Ingot casting	1600	MLT steel	2800	1.175	383	Wtr out	212	0.05	50
			Wtr in	75	0.05	0	Ingot	1600	0.59	107
			Electric			300	Ingot	75	0.56	0
							Ingot	75	0.002	0
							Ingot	75	0.023	0
							Loss			526
12	Soaking Pit	2400	Ingot	1600	0.59	107	Stack	1200	0.96	260
			Ingot	75	0.56	0	Ingot	2400	1.15	325
			Coke gas (Fuel)		0.04	750	Loss			280
			Air	75	0.66	0				
			Electric			10				

(Continued)

TABLE 4.5C Continued

No	Description	Temperature (°F)	Flow	Temperature (°F)	Mass (LB)	Energy (BTU)	Flow	Temperature (°F)	Mass (LB)	Energy (BTU)
13	Roughing mill	1900	Ingot	2400	1150	325	Blooms	1900	1.14	228
			Electric			110	Scale	1900	0.01	2
							Loss			205
14	Reheat furnace	2400	Slabs	75	0.54	0	Slabs	2400	0.41	113
			Fuel			750	Slabs	2400	0.13	36
			Air	75	1.4	0	Stack	1200	1.4	416
							Loss			185
15	Plate mill	1800	Slabs	2400	0.13	36	Plates	1800	0.13	25
			Electric			50	Loss			61
16	Hot mill	1800	Slabs	2400	0.41	113	Slabs	1800	0.36	68
			Electric			125	Scale	1800	0.05	9.5
							Loss			160.5
17	Pickle	150	Slabs	75	0.36	0	Slabs	100	0.36	9.9
			Acid – SLN	120	1	45	Cond	180	0.151	15.9
			Steam	320	0.151	187.7	CW	95	3.75	75
			CW	75	3.75	0	Acid- SLN	200	1	122
							Loss			10

ALUMINIUM SMELTING ENERGY INTENSITY (AC kWh/Ton)

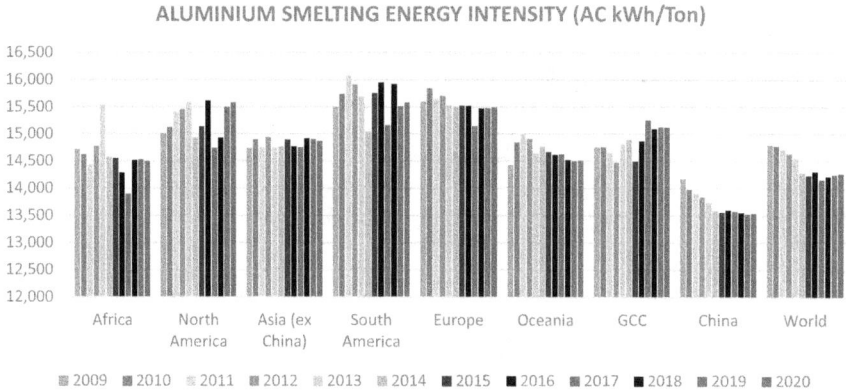

FIGURE 4.13 Primary aluminum smelting energy intensity (AC kWh/ton) trend For world regions [3].

TABLE 4.6
Primary Aluminum Smelting Energy Efficiency Trend for World Regions (Theoretical Limit 6230 kWh/Ton)

Year	Africa	North America	Asia (ex. China)	South America	Europe	Oceania	Gulf Countries	China	World
2009	42.3%	41.5%	42.3%	40.2%	39.9%	43.2%	42.2%	44.0%	42.1%
2010	42.6%	41.2%	41.8%	39.6%	39.3%	42.0%	42.2%	44.6%	42.2%
2011	43.2%	40.4%	42.2%	38.8%	39.8%	41.5%	42.5%	44.8%	42.3%
2012	42.2%	40.3%	41.7%	39.2%	39.7%	41.8%	43.0%	45.0%	42.6%
2013	40.1%	40.0%	42.2%	39.7%	40.1%	42.5%	42.0%	45.3%	42.8%
2014	42.8%	41.7%	42.2%	41.4%	40.2%	42.2%	41.8%	45.8%	43.6%
2015	42.8%	41.2%	41.8%	39.6%	40.1%	42.5%	43.0%	45.9%	43.8%
2016	43.6%	39.9%	42.2%	39.1%	40.2%	42.6%	41.9%	45.8%	43.6%
2017	44.8%	42.3%	42.2%	41.1%	41.1%	42.6%	40.8%	45.9%	44.0%
2018	42.9%	41.7%	41.8%	39.1%	40.3%	42.9%	41.3%	46.0%	43.8%
2019	42.9%	40.2%	41.8%	40.2%	40.3%	43.0%	41.2%	46.0%	43.7%
2020	43.0%	40.0%	41.9%	40.0%	40.2%	42.9%	41.2%	46.0%	43.6%

Source: Website of the International Aluminium Institute (www.world-aluminium.org/statistics/primary-aluminium-smelting-energy-intensity/, accessed on August 13, 2021)

Obaidat, Mazin, et al. Energy and Exergy Analyses of Different Aluminum Reduction Technologies; Sustainability 2018, 10, 1216; doi:10.3390/su10041216

A typical energy and mass flow diagram for aluminum manufacturing comprising Bayer process and Hall–Héroult process is shown in Figure 4.14.

4.5.3 ENERGY REQUIREMENTS IN COPPER PRODUCTION

Typical energy and mass flow in copper manufacturing plant are shown in Figure 4.15 and Table 4.7.

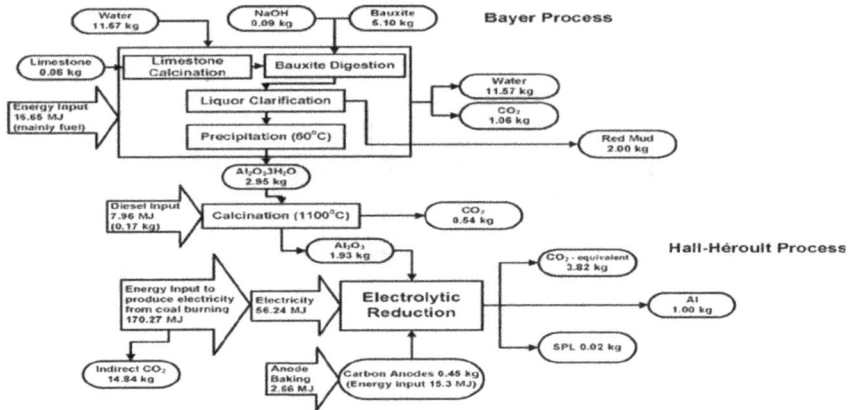

FIGURE 4.14   Mass and energy balance of aluminum manufacturing.

Source: www.researchgate.net/publication/234004241_Energy_and_Exergy_Analysis_of_the_Primary_Aluminum_Production_Processes_A_Review_on_Current_and_Future_Sustainability/figures?lo=1

4.6 CONCLUSIONS WITH SOME ENERGY-SAVING POTENTIAL IN METAL INDUSTRIES

4.6.1 ENERGY CONSUMPTION PATTERNS IN A FORGING INDUSTRY

Heating and heat treatment furnaces are the major thermal and electrical energy consumers in the forging industry. Process heating/furnaces alone account for 50–80% of the total energy consumption. Electrical motors and compressed air systems account for 15–20%, and others like lighting, machining, grinding, shot-blasting, etc., account for 10–15% of total energy consumption in a forging plant.

4.6.2 ENERGY CONSERVATION MEASURES IN A FORGING INDUSTRY

In the forging industry, substantial reduction in energy consumption can be achieved by improving the operational practices, fine tuning the operating parameters, application of low-cost automation and upgrading technology. A list of possible energy conservation opportunities in three categories are short-term, medium-term and long-term energy savings projects.

4.6.3 THREE TYPES OF ENERGY-SAVING POTENTIAL IN METAL INDUSTRIES

The energy-saving potential considering the short-term and medium-term energy-saving projects is 10–15% of the total energy consumption. The energy-saving potential considering the long-term energy-saving projects, which have payback period of about 3–4 years, is in the range of 15–20%. Some of such measures for iron and steel industry in particular are shown in Figure 4.16 and for metal industries in general are shown in Chapter 5.

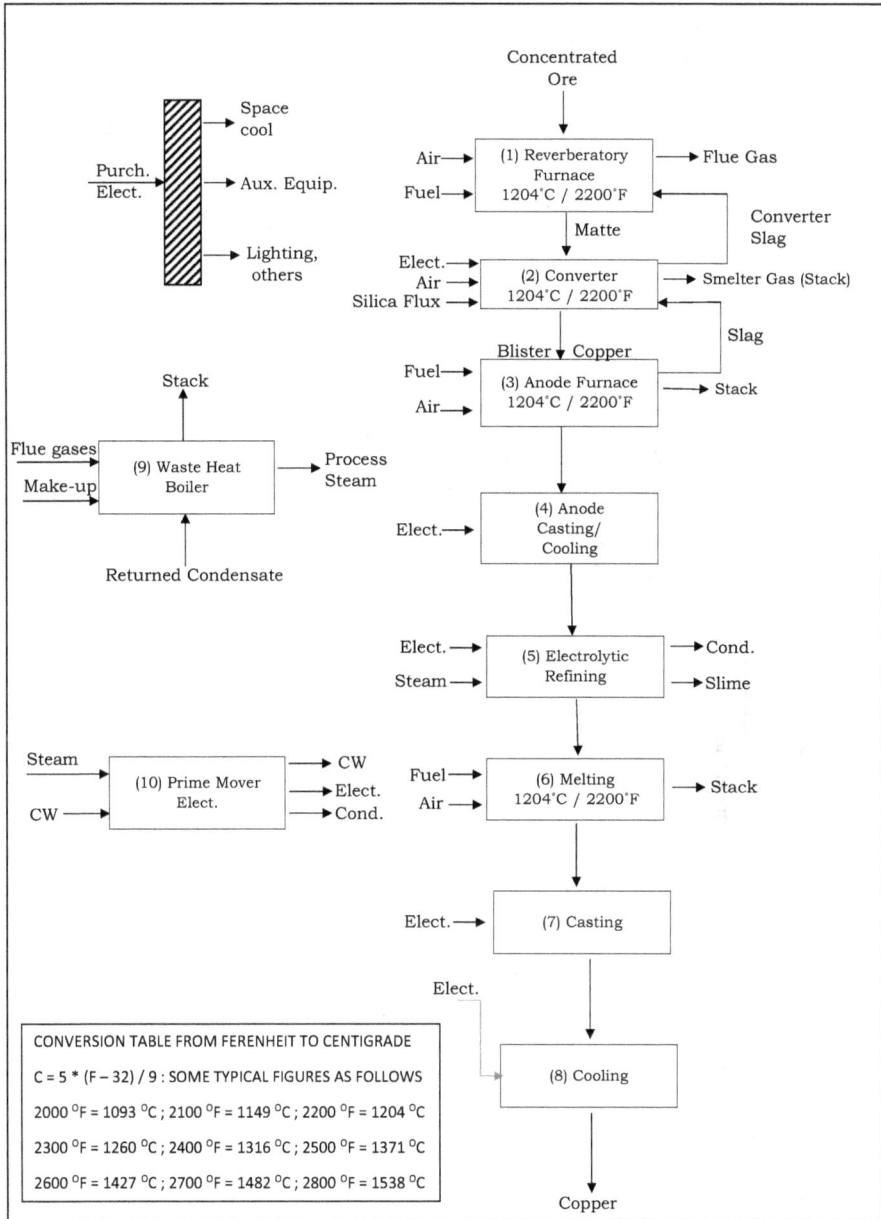

FIGURE 4.15 Typical energy and mass flow in a copper manufacturing plant.

Source: Brown, Harry L et al. (1985) Energy Analysis Of 108 Industrial Processes, Drexel University Philadelphia. PA 19104, US Department of Energy

TABLE 4.7

Typical Energy and Mass Flow in a Copper Manufacturing Plant

No	Description	Temperature (°F)	Inlet				Outlet			
			Flow	Temperature (°F)	Mass (LB)	Energy (BTU)	Flow	Temperature (°F)	Mass (LB)	Energy (BTU)
1	Reverberatory furnace	2200	Ore	75	4.12	0	MATTE	2200	3.4	1439
			Limestone	75	0.2	0	FLUE GAS	1500	20.13	7113.6
			Con V SLG	2200	3.08	1319	SLAG	2200	4	1700
			Air	75	20.13	0	LOSS			2806.4
			Fuel			11750.3				
2	Converter	2200	Matte	2200	4	445	Copper	2200	1.02	323.9
			Air	75	3.67	0	Salt gas	2000	3.67	1710.6
			Slag	2200	0.01	4.3	Slag	2200	3	1309
			Silca	75	0.72	0	Loss			3834.6
			Electric			235				
			Exotherm			5500				
3	Anode Furnace	2200	Copper	2200	1.02	323.9	Copper	2200	1.01	314.2
			Air	75	1.3	0	Slag	2200	0.01	4.3
			Fuel			1750	Stack	1400	1.3	490
							Loss			1265.4
4	Note Casting	2200	Copper	2200	1.01	314.1	Anodes	75	1.01	0
			Electric			380	Loss			694.2
5	Electrolytic refine	140	Anodes	75	1.01	0	Cathodes	140	1	6
			Steam	250	0.13	150	Cond	180	0.13	20
			Electric			475	Slime	180	0.01	0.5
							Loss			598.5
6	Melting	2200	Cathodes	75	1	0	Copper	2200	1	280
			Air	75	3.46	0	Stack	1100	3.46	1100
			Fuel			1500	Loss			140

No	Process	Input				Output			
7	Casting	Copper	2200	1	280	Copper	2200	1	280
		Electric	2200	1	380	Loss			380
8	Cooling	Castings	75	1	280	Castings	2200	1	0
						Loss			280
9	Boiler	Flue gas	1500	20.13	7113.6	Steam	250	2.6	2865
		Cond	180	2.3	241.5	Steam	567	0.74	850
		Make up	75	1.04	0	Stack	700	20.13	2760
						Loss			880
10	Electric generation	Steam	567	0.74	850	CL wtr	100	17.44	436
		CL wtr	75	17.44	0	Cond	180	0.74	74
						Elect			300
						Loss			40

Source: Brown, Harry L et al. (1985) Energy Analysis Of 108 Industrial Processes, Drexel University Philadelphia. PA 19104, US Department of Energy

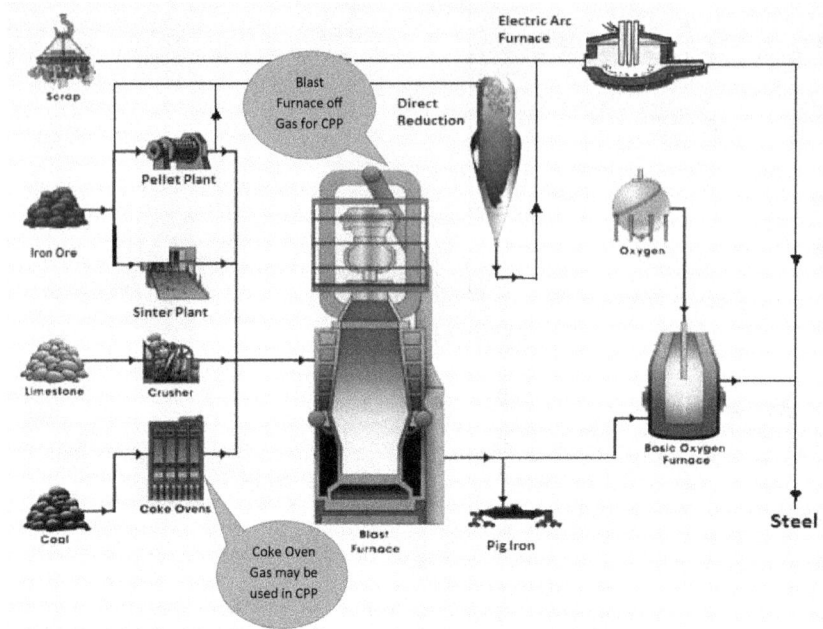

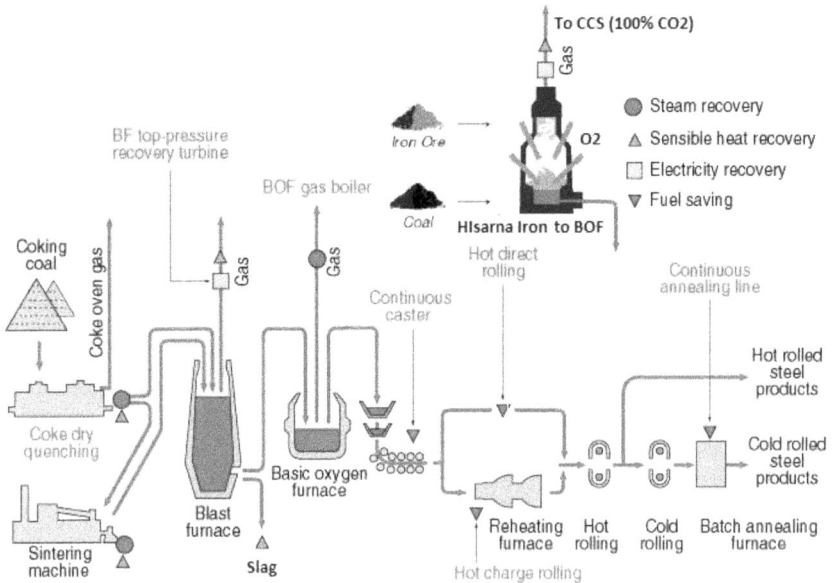

FIGURE 4.16 Energy-saving potential in steel works.

Recycling metals has the multiple advantages of materials and energy conservation. The energy saving achieved through use of recycled metal products after their end of useful life to produce secondary metals, instead of using ore to produce

TABLE 4.8

Ranges of Energy Savings through Recycling for Various Ferrous and Non-Ferrous Metals

Metal/Product	% Savings
Aluminum	90–97
Copper	84–88
Gold	98
Lead	55–65
Magnesium	97
Nickel	90
Palladium	92–98
Platinum	95
Rhodium	98
Silver	96
Steel	60–75
Stainless steel (304)	68
Titanium	67
Zinc	60–75

Source: UNEP (2013)

primary metals, has been estimated and shown in Table 4.8. Society needs to attain 100% recycling, where applicable, as this initiative would not only save up to 98% energy of manufacturing primary metals, but also help establish the circular economy, benefiting more people in the process.

Thus, EE&C is essential for sustenance of metal industries and could be achieved through both tact and tactics of appropriate use of technology and techniques in the production process, taking into consideration the entire life cycle of metals.

REFERENCES

BEE (2015); *Energy Performance Assessment for utility and Equipment Systems, Guide Book for National Certification Examinations for Certified Energy Manager and Certified Energy Auditor*, Bureau of Energy Efficiency (BEE), Government of India, 2015.

CRU International (2018); The Role of Silver in the Green Revolution, CRU International prepared for The Silver Institute, July 2018

Energypost (2020); *HYBRIT Project Sweden Goes for Zero-Carbon Steel*, 2020; https://energypost.eu/hybrit-project-sweden-goes-for-zero-carbon-steel/

IEA (2020); *Iron and Steel Technology Roadmap*, IEA, 2020; https://www.iea.org/reports/iron-and-steel-technology-roadmap

PMI (2018); *Hisarna – Developing a Sustainable Steel Production Process at Tata Steel Europe*, PMI, 2018; www.pmi.org/-/media/pmi/documents/public

Schmale, K., Kersten, D., and CONARC® (2003); *All Purpose Steelmaking by SMS Demag*, 2003; http://metal2014.tanger.cz/files/proceedings/metal_03/papers/185.pdf

Steel online Open Energy Monitor (2021); https://learn.openenergymonitor.org/sustainable-energy/energy/industry-steel

UNDP (2013); Environmental Risks and Challenges of Anthropogenic Metals Flows and Cycles; Report #3 of the Global Metal Flows Working Group of the International Resource Panel of UNEP; 2013.

WEBSITES

[1]   https://en.wikipedia.org/wiki/Silver
[2]   www.steelguru.com/steel/progress-update-on-3-hybrit-pilot-projects
[3]   Website of the International Aluminium Institute (www.world-aluminium.org/statistics/primary-aluminium-smelting-energy-intensity/, accessed on August 13, 2021)

5 Energy Audits in Metal Industries

Swapan Kumar Dutta

5.1 INTRODUCTION

Metal industries have been identified as being energy intensive and having a significant share of energy consumed among industry sector. Appropriate policy provides the necessary guidelines and legal frameworks to attain higher efficiency and minimize waste. Energy audit is the necessary tool to identify the techno-economically viable projects so that implementation of the same will yield the intended benefits. Thus, energy audit plays a most important role in day-to-day operation of metal industries aspiring to improve energy efficiency and conservation (Jarza, 2011). Therefore, a handy overview and various energy audit procedures is discussed in this chapter.

5.2 ENERGY AUDITS IN METAL INDUSTRIES

The definition of energy audit as defined under the Energy Conservation Act of 2001 (BEE, 2022) is as follows: "The verification, monitoring and analysis of the use of energy and submission of technical report containing recommendations for improving energy efficiency with cost benefit analysis and an action plan to reduce energy consumption."

In metal industries, energy audits can be simply categorized into two types, viz. preliminary energy audits and investment grade energy audits.

5.3 PRELIMINARY ENERGY AUDITS

A preliminary energy audit generally requires relatively short time of plant visit. It focuses on readily available macro data provided by the unit/plant, or data/information collected during the preliminary audit. The objectives of a preliminary energy audit are as follows:

1. Familiarization of metal production process plant activities.
2. Observation and assessment of current level operation and practices.
3. Identifying energy-saving potential areas in the plant.
4. Identifying no-cost or low-cost areas for energy savings with immediate results.
5. Identifying and shortlisting potential areas for more detailed study or assessment.

DOI: 10.1201/9781003157137-5

5.4 INVESTMENT GRADE ENERGY AUDITS (IGEA), A.K.A. DETAIL ENERGY AUDITS

An investment grade energy audit, also known as a detailed energy audit, is a process of accounting and analyzing the present energy consumption to identify potential energy-saving possibilities. An investment grade detailed audit report contains recommendations for improving energy efficiency with cost/benefit analysis, and technical specifications for any retrofit options with the list of suppliers and manufacturers of energy-efficient technologies.

An investment grade energy audit has the following features:

1. Detailed energy balance of the plant.
2. Energy consumption pattern analysis of energy source mix and percentage share of each.
3. Performance assessment of all energy consuming equipment, along with auxiliary units.
4. Energy efficiency measures (EEMs) with cost/benefit analysis.
5. Investment analysis of proposed EEMs (includes estimation of internal rate of return [IRR], net present value [NPV] and payback).
6. Required action plan for implementing EEMs.

5.5 VIRTUAL ENERGY AUDITS

In recent years, with the advent of the IoT, a new energy audit technique has been developed and offered by many energy audit service providers and is often termed as a virtual energy audit, whereby much of the energy audit activities could be conducted remotely with no or minimum physical intervention of the energy audit team

---

TABLE 5.1

Structure Format of Investment Grade Detailed Project Report (IGDPR)— Energy Efficiency Measures Summary Format

EEM NO.	Description of EEM	Power Savings (KWh/yr)	Peak Demand Reduction (KW)	Fuel Savings (Units)	Energy Cost Savings (Rs/yr)	Estimated Investment (Rs)	Estimated Life (Yrs)	Payback	IRR and/or RoI

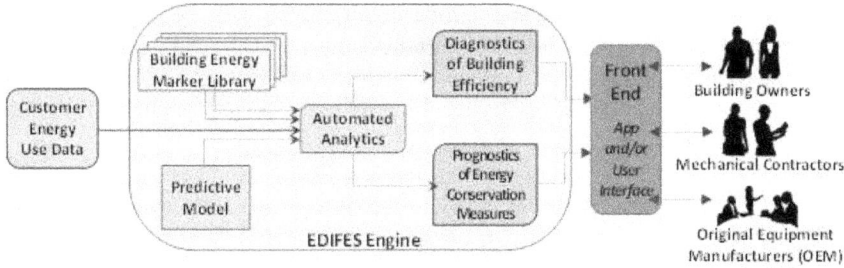

FIGURE 5.1   Virtual energy audit.

at the facility site to be audited. This has become possible because of technological breakthroughs in remote monitoring of energy gadgets in plants/buildings and machinery. Where applicable and when properly implemented, virtual energy audit procedure could be much cost effective, faster and convenient.

Virtual energy audit (Avina and Rottmayer, 2016) is fast becoming an option in place of physical audit for front-end cost advantage, less involvement of manpower, more easy to understand and more suitable for repetitive activities, such as, rechecking/re-doing the various steps of virtual energy audit. However, each of these advantages could turn out to be potential disadvantage as physical audit could be, in many cases, ultimately less expensive (such as, if virtual audit leads to investment grade audit and preparation of detailed project report [DPR], which involves additional cost).

Virtual audit may be cause of lost manpower (as faults in virtual audit process when detected may be far from corrective action), and a too-simple and contemporary approach may not find the real opportunities and innovative solutions.

5.6   PROCESS SYNTHESIS-MINIMUM ENERGY APPROACH

In any metal industry when an energy audit is conducted, the whole process—including auxiliaries—should be analyzed for minimum energy consumption with maximum production. Electrical and thermal energy should be used properly, keeping an eye on minimum energy approach to achieve minimum cost to the company (IBEF, 2009). This can be achieved through one or combination of the following:

1. Process selection from alternatives considering:
    i. Availability/cost of raw materials required.
    ii. Cost of equipment purchase and maintenance.
    iii. Cost of thermal and electrical energy used.
    iv. Availability/cost of manpower required.
2. Minimization of operating cost of energy guzzlers which calls for:
    i. Its proper maintenance.
    ii. Its proper control, for example, use of variable frequency drive, if applicable.
    iii. Its timely replacement with a more efficient version.

3. Negotiating and adopting an appropriate tariff plan with the utility or finding optimum operation practice with the captive power plant through:
   i. Optimizing load factor and power factor.
   ii. Avoiding peaking of energy use by production scheduling, use of maximum demand controller, etc.

For example, proper use of tariffs such as time of the day (ToD), availability-based tariff (ABT), etc., through detailed discussion and negotiation with the utility in electricity helps in reaching goals in minimum energy approach fundamentals. Production schedule should be at peak when the electricity cost is minimum (in off peak period), and it should be minimum when the electricity rate is high (peak period in the day). Proper planning of production schedule helps in achieving a substantial amount of power savings cost per year. Consumption parameters are typically recorded at 15-minute intervals. Precautions should be taken to avoid exceeding the contract demand for more than the recording interval period, which may invite penalty—perhaps even for a prolonged period—if the ratchet clause is in place. Understanding the terms and conditions of the utility helps to select appropriate tariff plan and keep the electricity bill to a minimum without hampering the production.

The SEC or the energy consumed per unit mass is an indicator of the performance of any equipment. The lower the SEC, the better is its energy performance (BEE, 2018).

Energy audit takes care of all the previously mentioned issues.

5.7 TOOLS FOR ENERGY AUDITS

The performance assessment of utilities depends on the availability of reliable and accurate data from individual equipment used in the production, its processing and its presentation in an appropriate format to help analysis and find opportunities for improvement, made possible through use of several tools for energy audit, such as measuring instruments, its instrumentation and software.

The electrical control panels installed in the plant provides data related to production, energy consumption and key operational parameters. Additional data that would be required for the purpose of performance assessment is collected using portable instruments. Some of the portable instruments commonly used in energy audits are: power analyzer, ultrasonic flow meter, flue gas analyzer, hygrometer, digital temperature indicator, thermal imager, lux meter, infrared thermometer, anemometer, temperature data logger, stroboscope, etc. (Figure 5.2).

5.8 INSTRUMENTATION

The entire metal manufacturing/processing system has to be controlled properly to achieve more economical, efficient and reliable operations in real time. Instrumentation is the process of controlling, measuring and analyzing physical quantities using various types of interconnected process control instruments. Just like information run through appropriate processes can become useful data, so the sensors of previously mentioned instruments and various other types of instrumentation

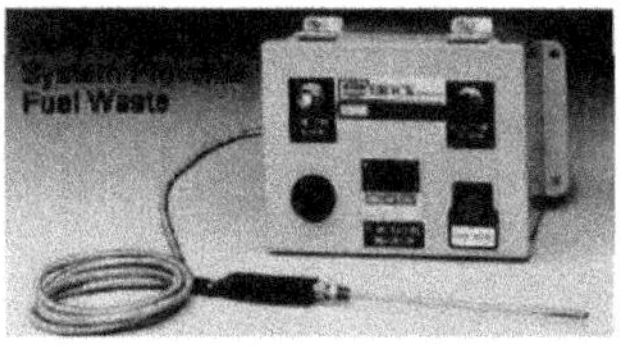	**Fuel Efficiency Monitor:** This measures Oxygen and temperature of the flue gas. Calorific values of common fuels are fed into the microprocessor which calculates the combustion efficiency.
	**Combustion Gas Analyzer:** This instrument has in-built chemical cells which measure various gases such as CO_2, CO, NO_X, SO_X etc. Gas analyzers are flexible in what must be measured depending on the requirements of the customer/user. They have specific sensors sealed inside the equipment that can be changed to measure the different components in the gas. But because a maximum of two sensors can be connected, only two or three parameters can be measured at one time It is light and easier to handle compared to the fuel efficiency monitor.
	**Manometer with Pitot Tube:** Digital flexible membrane manometer is used for measuring pressures in air ducts carrying exhaust flue gases (boiler, furnaces), or air from fans and blowers. • To measure pressure in air pipes, manometers must be used in combination with a pitot tube • Attach flexible rubber tubes to the inlet and outlet probes of the manometer. Tighten these to ensure that there is no leakage of air. • Attach these two tubes to the ends of the pitot tube • Make a 6-cm monitoring hole in the duct or pipeline • Insert the pitot tube into the monitoring hole

FIGURE 5.2   Tools for energy audits.

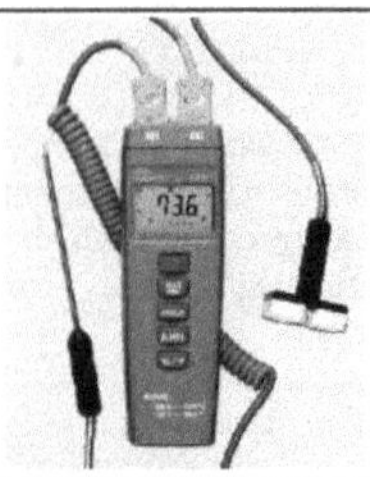	**Contact Thermometer:** These are thermocouples which measures for example flue gas, hot air, hot water temperatures by insertion of probe into the stream. For surface temperature a leaf type probe is used with the same instrument.
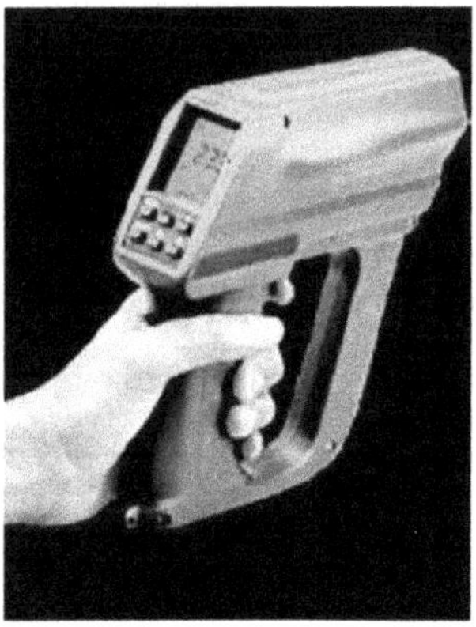	**Non Contact Infrared Thermometer:** Infrared thermometers calculate the amount of thermal radiation (infrared radiation) emitted from the object. By knowing the emissivity of the object and the amount of infrared energy emitted by the object, the object's temperature can be determined. With the help of infrared thermometers, temperatures of the objects placed in hazardous or hard-to-reach places or other situations can be determined. The most common design of a IR thermometer consists of a lens to focus the infrared energy on to a detector. The detector changes the energy to an electrical signal that can be shown in units of temperature after being corrected for ambient temperature variation.
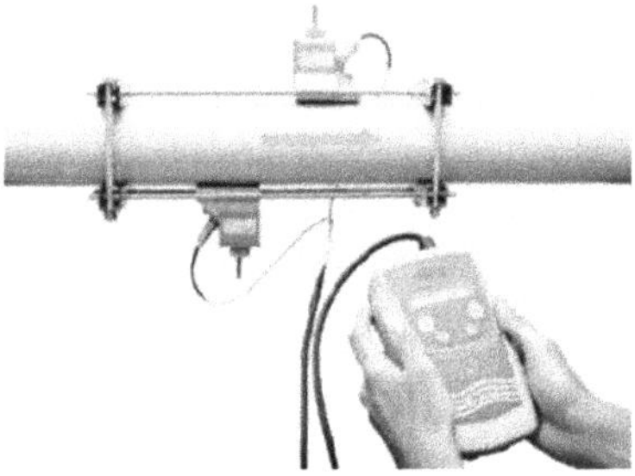	**Ultrasonic Flow Meter:** This is one of the popular means of non-contact flow measurement. There are two main types of ultrasonic flowmeters: Transit time and Doppler. Transit time ultrasonic meters have both a sender and a receiver. They send two ultrasonic signals across a pipe: one traveling with the flow and one traveling against the flow. The ultrasonic signal traveling with the flow travels faster than a signal traveling against the flow. The ultrasonic flowmeter measures the transit time of both signals. The difference between these two timings is proportional to flow rate. Transit time ultrasonic flowmeters usually monitor clean liquids. Doppler ultrasonic flowmeters measure dirty liquids. They compute flow rate based on a frequency shift that occurs when their ultrasonic signals reflect off particles in the flow stream.

FIGURE 5.2   (Continued)

		Speed Measurements:
		In any audit exercise speed measurements are critical as they may change with factors such as frequency, belt slip, loading, etc A simple tachometer is a contact type instrument, which can be used where direct access is possible.
		More sophisticated and safer ones are non contact instruments such as stroboscopes. A stroboscopic light source provides high-intensity flashes of light, which can be caused to occur at a precise frequency. When this light source is made to fall on an object with periodic motion it appears that the motion is slowed down, or stopped when both the frequencies bear a definite relationship. A stroboscope uses this Principle for measurement of RPM.
Tachometer	Stroboscope	
		Psychrometer: A sling psychrometer - consists of two thermometers mounted together with a handle. One thermometer is ordinary and measures the dry bulb temperature. The other has a wet cloth wick, over its bulb and is called a wet-bulb thermometer. When a reading is to be taken the psychrometer is whirled around. The water evaporates from the wick, cooling the wet-bulb thermometer. Then the temperatures of both thermometers are read. If the surrounding air is dry, more moisture evaporates from the wick, cooling the wet-bulb thermometer more, so there is a greater difference between the temperatures of the two thermometers. By using these temperatures the humidity is computed.

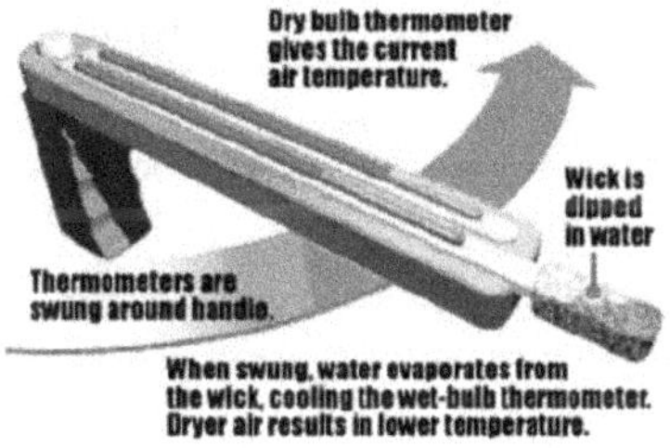

FIGURE 5.2   (Continued)

	Lux Meters: A light sensitive cell measures the incident light (all light in the visible spectrum is measured) and evaluates that against the human daylight sensitivity curve. The resulting value is the measurement result in lux. This works well but it requires a different correction factor for every light spectrum. The much more expensive lux meters with one cell are optimized and tuned with optical filters and lenses such that the sensitivity of this set of lenses and the cell itself directly matches the eye's light sensitivity curve (so only one correction value needed for light of any spectral content).
	**Smart Energy Meters:** The term smart meter usually refers to electric meters which keep detailed statistics on usage, but it can be used for fuels or water applications as well performing the same job. The primary purpose of smart meters is to provide information on how end users use their electricity on a real-time basis. The smart energy meters use a wireless communication to help track the electricity consumption and thus save both electricity and money. It can be easily installed and gives an accurate reading of electricity consumption which can also be monitored / controlled through mobile or internet.
	**Thermography** Infra-red thermal monitoring and imaging (non-contact type) measures thermal energy radiation from hot/cold surfaces of an object and provides input for assessing health of equipment and predictive maintenance. The thermal camera unit converts electromagnetic thermal energy (IR) radiated from an object into electronic video signals. These signals are amplified and transmitted via interconnected cable to a display monitor where the resulting image is analysed and interpreted for hot/cold spots.

FIGURE 5.2　(Continued)

are being used to monitor and maintain process control equipment. For controlling any quantity as previously mentioned, primarily that particular quantity has to be measured, for improved production, product consistency and quality management, and workplace safety, preventing fire or explosion in manufacturing and processing facilities. Industrial automation and industrial instrumentation are thus required to control various operations in industries. As most of the metal industries are automated, using embedded systems for ease of operation, the microcontroller, microprocessors and computer are required for programmable switching.

5.9   SOFTWARE APPLICATIONS

Both for analysis of data collected from instruments and also for automation/automatic control of metal industry equipment for process control, appropriate software must be in place, analogous to the operating system which is in place to run the computers. Thus, software is an indispensable component of any EE&C activity of metal industries. Many of the useful software products are available free, while their commercial version or others are costly, but appropriate application makes them cost effective. Some of such software products are listed in what follows.

- Energy audit software (e.g. MotorMaster+ Tools, AIRMaster+, Measure, etc.).
- Energy accounting, billing and trend analysis software (e.g. spreadsheets, EnergyStar portfolio manager, Labs 21 for laboratories and commercial versions such as EnergyCap, Energy Watchdog Pro, Matrix, etc.).
- Energy performance simulation software (e.g. DOE-2, eQuest, EnergyPlus, TRNSYS and commercial versions of these, as well as TRACE of TRANE, HAP of Carrier, etc.).
- Other software (e.g. US DoE, LBNL Tools).

Some sources of software:

- www.autodesk.com/
- https://energyplus.net/
- www.energy.gov/eere/amo/software-tools
- www.energy.gov/eere/amo/downloads/motormaster-tool
- www.energy.gov/eere/amo/measur
- 50001 Ready Navigator (https://navigator.lbl.gov/)
- www.green-buildings.com/articles/free-energy-audit-software-available/
- www.energyauditsoftware.com/download.htm
- http://poet.lbl.gov/cal-arch/links.html
- Building Life Cycle Cost including Renewble Energy: http://www1.eere.energy.gov/femp/information/download_blcc.html
- Building specific software at http://apps1.eere.energy.gov/buildings/tools_directory/doe_sponsored.cfm
- MATLAB (https://matlab.en.softonic.com/)
- Others at http://www1.eere.energy.gov/analysis/

Training is often necessary for the appropriate team member(s) for appropriate software to enable an energy management team to achieve EE&C in metal industries.

5.10   ENERGY AUDITS IN FERROUS AND NON-FERROUS INDUSTRIES

Energy audit is a key requirement for any metal industry, including ferrous and non-ferrous steel plants (FICCI, 2018). It helps to identify energy efficiency improvements in a systematic way. It assists the metal industry management in understanding how to use energy efficiently and helps to identify the areas where energy waste occurs and where opportunities for improvement (OfIs) exist.

Energy audit is the inspection, examination, analysis and evaluation of the physical and financial processes of the metal industry relating to the use of the energy. The aim of the energy audit is to systematically identify the potential for energy savings and make recommendations for improvement. Hence, the energy audit is an effective measure to realize the energy savings for the steel plant. It also creates a platform for the management to compare its performance with other steel plants for performance improvement (Teri, 2015).

The objective of conducting an energy audit can vary from one metal industry to another, but the main objective of reducing the use of energy remains the same. Energy audits are also conducted to evaluate the effectiveness of an energy efficiency program (BEE, 2016).

The energy audit of a metal industry consists of the following activities.

REQUIREMENTS OF THE ENERGY AUDIT

1. Energy management processes.
2. Energy consumption in the process.
3. Energy management and statistical data.
4. Analysis of the energy consumption using the data.
5. Analysis of the comprehensive energy consumption of products and output values.
6. Analysis of energy costs.
7. Calculation of energy savings.
8. Financial and economic analysis of proposed energy-saving technologies.
9. Identifying potential suppliers and service providers of the proposed energy-saving technologies.

5.11   PERFORMANCE ACHIEVEMENTS AND TRADE (PAT) CYCLES OF INDIA-IRON AND STEEL PROJECTIONS AND ACHIEVEMENTS UNTIL 2030

Iron and Steel Projections and Achievements until 2030	Units	Values
Number of DCs in the sector	nos	67
Total energy consumption of DCs in the sector	million TOE	25.32
Total energy saving target for iron and steel sector in PAT Cycle-I	million TOE	1.486
Total energy savings achieved by iron and steel sector in PAT Cycle-I	million TOE	2.1
Energy savings achieved in excess of the target	million TOE	0.614
Reduction in GHG emissions in Cycle-I	million T CO2	6.51
Cumulative energy savings with impact of PAT till 2030 over BAU (business as usual) consumption	million TOE	29.88

Methodology followed for impact assessment of PAT Cycle-1 is explained in what follows.

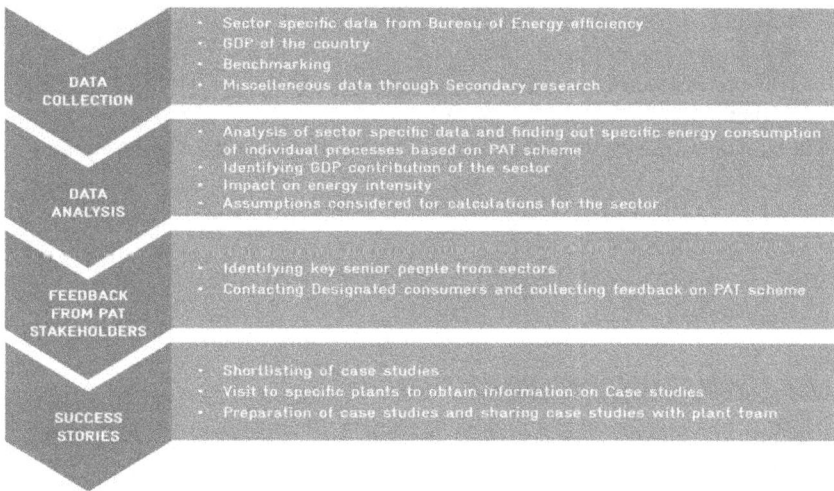

5.12   ENERGY-SAVING POTENTIAL IN INDIAN METAL INDUSTRIES

The energy-saving potential considering the short-term and medium-term energy-saving projects is 10–15% of the total energy consumption and have typically simple payback periods of less than one year and 1–3 years, respectively. The energy-saving potential considering the long-term energy-saving projects, which have payback period of about 3–4 years, is in the range of 15–20%.

5.12.1 SHORT-TERM ENERGY-SAVING PROPOSALS

Better planning of job schedules for optimization of the overall loading of furnaces.

Heating and heat treatment furnace combustion efficiency improvement.

Furnace operation at optimum temperature.

Close furnace openings to control heat losses.

Maintenance of proper amount of furnace draft.

Periodic maintenance to arrest compressed air leakages.

Compressor operating pressure optimization based on the system requirement.

Lightly loaded motors' delta connections to be replaced with permanent magnets.

5.12.2 MEDIUM-TERM ENERGY-SAVING PROPOSALS

IFs and other larger electrical machines should be installed with kWh meters.

Maximize the use of day light by use of transparent sheets for rooftop of workshop/production facilities and window panes of the buildings (ACEEE, 2021).

The plant for hot air exhaust to be fitted with eco ventilators.

Forging and heat treatment furnace to employ automation for temperature control.

Overall insulation levels to be improved.

Improve power factor by installation of automatic power factor control (APFC) System.

The lightly loaded motors which handle fluctuating loads need to be equipped with automatic soft starters.

Install automatic voltage stabilizers for lighting circuits and other precision electronic circuits.

Install lighting transformers in all major lighting feeders and operate the lighting circuit at 220 V.

5.12.3 LONG-TERM ENERGY-SAVING PROPOSALS

The heat treatment furnaces and heating furnace to be installed with air preheater for pre-heating the combustion air supply.

Existing oil-fired heating and heat treatment furnaces to be replaced with gas-fired furnaces.

Back operated furnace provided with ceramic fiber insulation.

Practice oxygen level increase in reheating furnace.

In metal industries, segregate high-pressure and low-pressure compressed air equipment.

For the compressed air system and places with variable loads, employ adjustable or variable-speed drives.

Save energy by replacing pneumatic-operated tools with electrical tools.

Install a energy-efficient reheating furnace with a flat roof.

Use energy-efficient motors replacing old motors.

5.13   CASE STUDIES FOR CLASSIFIED METAL INDUSTRIES

Case Study 5.1: Direct Charging of Hot Billet from CCM to Roll Mill

INTRODUCTION

Shri Bajrang Power & Ispat Ltd. (Borjhara Division), a Goel Group Company in India, has the following.

Total plant capacity	2,10,000 TPA
Plant divisions	2 × 350 TPD
Captive power plant	18 MW
Induction furnace	5 × 8 MT
Continuous casting machine	1 × 29,000 TPA
Biomass power plant	8 MW
Ferro alloys plant	14,000 TPA
Coal washery	0.60 MMT
Rolling mill	1.20 LTPA
Fly ash bricks	30 million bricks PA

PROJECT BACKGROUND

The unit first used the conventional top-fired pusher type reheating furnace to heat the billets. The rising price of fuel being used, like furnace oil with provision for coal-based producing gas, was leading to higher production cost. The comparison of conventional process & modified process is given in Figure 5.3.

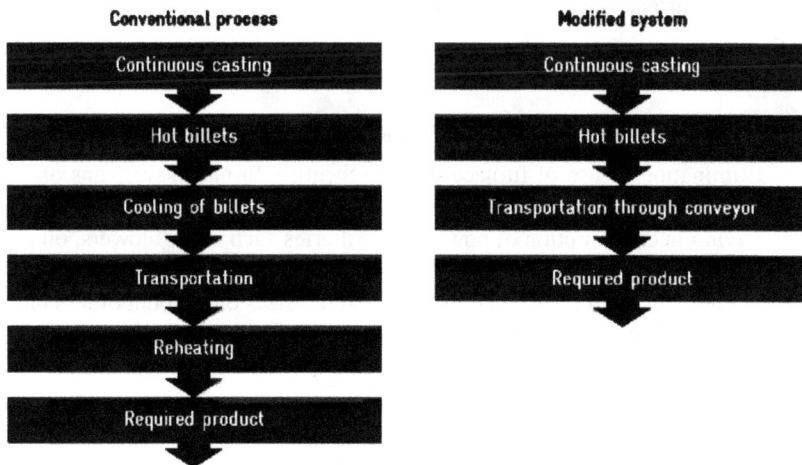

FIGURE 5.3   Comparison of conventional process and modified process.

A direct rolling of hot continuous-cast billet for producing TMT bars was one of the major energy-efficient measures undertaken by the company. Under this technology, at Borjhara division, a rerolling mill of 1.2 LTPA capacity was installed, replacing the use of a reheating furnace. As a result, the fuel costs, refractory costs and running costs of reheating furnace were eliminated.

DESCRIPTION OF THE PROJECT

The installation of direct rolling of hot continuous-cast billet without any intermediaries has led to energy and environmental conservation. The requirements for upgrading the unit are as follow.

CHANGES AT THE INDUCTION FURNACE

The tapping temperature of the molten steel has been increased, and this high temperature is retained by modification of the ladle and its transfer mechanism. Moreover, the operation of IFs is synchronized with CCM and rolling mill to enhance mill utilization.

CHANGES AT CONTINUOUS CASTING MACHINE

The speed of casting, temperature, water pressure, cooling and water circulating system is modified. The temperature of cast billet is set at 1080–1100°C. The secondary cooing system is achieved by a PLC-based automation system. Manual gas cutting is replaced by hydraulic shearing, and high-speed billet conveyor is installed with VFD drives.

CHANGES AT THE ROLLING MILL

The roller pass design (RPD) is modified to improvise the direct rolling. Reduction of idling of the rolling mill is achieved by using two-strand casting and increasing input feed.

CHANGES AT THE FINISHED MATERIAL YARD

A 5S system (sort, set in order, shine, standardize and sustain) has been implemented to carry the material handling in systematized way.

ACHIEVING REDUCED SPECIFIC ENERGY CONSUMPTION

- Elimination of use of furnace oil for reheating furnace saves tons of fuel annually.
- Saving in consumption of power of auxiliaries such as air blowers, oil preheater and pumps, and pusher and ejector with associated conveyors.
- Higher yield percentage due to reduction in losses on account of less burning/scale loss and reduction in misrolled and end cuts, as well.

Duration of the project	Six months
Energy savings	Rs. 146 million
Annual TOE savings	4,100 TOE
Investment	Rs. 22 million

Payback	Two months
GHG reduction	13,000 tons of CO_2
Shutdown required	One month
Return on investment	663 %

Case Study 5.2: Reheating Furnace with Regenerative Burner at Tata Steel

INTRODUCTION

Tata Steel is a multinational steel making company, a subsidiary of Tata Group. In FY-17, with 27.5 MMT of crude steel deliveries, it became one of the leading steel producers globally and 2nd largest in India with 13 MMT of capacity.

PROJECT BACKGROUND

At Tata Steel in India, there had been a surplus of byproduct gases, generally of blast furnace (BF) gases, due to various forms of capacity expansion. These BF gases have calorific value of about 800–850 kCal/Nm³, and around 70–80 kNm³/hr is used to get flared, leading to energy losses of about 63 GCal/hr.

In 2008–2009, there was a requirement of reheating furnace due to capacity expansion from 2.9 MTPA to 3.55 MTPA, along with a supply of rich gases of more than 2,000 kCal/Nm³ of CV in combination with preheated air up to 500°C through the recuperator.

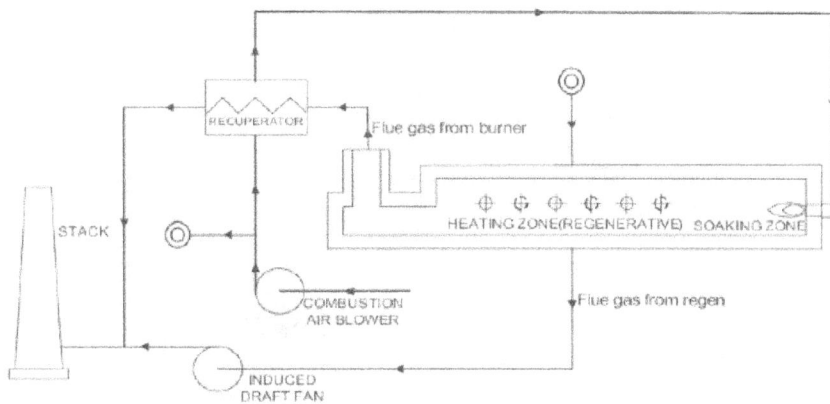

FIGURE 5.4   A layout of reheating furnace with regenerative burners.

DESCRIPTION OF THE PROJECT

Use of regenerative burners (Photo 5.1 and Photo 5.2) make it possible to utilize the surpus BF gas to heat slabs. It can use low-CV gas, i.e. BF gas with pre-heating provision up to 1,000°C. From 30–40% of fuel savings and required rolling temperature of slabs can also be achieved with this technology.

Out of 108 burners, 56 nozzles are located at air side and 56 nozzles are located at gas side. Each air nozzle and adjacent gas nozzle combines to form one combustion unit. The furnace is equipped with three-way reversing valve to help the combustion process and fume exhaust gas process.

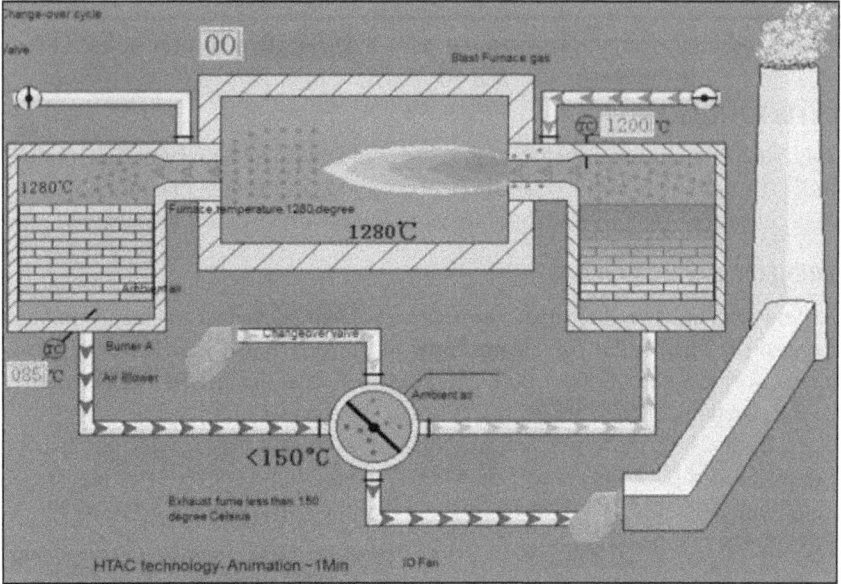

PHOTO 5.1   The operating principle of regenerative burners.

PHOTO 5.2   Regenerative burners.

BENEFITS

Annual energy savings	Rs. 110 million
Annual TOE savings	2,800 TOE
Investment	Rs. 200 million
Payback	22 months
GHG reduction	170,000 tons of CO_2
Return on investment	55%
Rapid heat transfer and improved temperature efficiency.	

Case Study 5.3: Scrap Pre-Heating of Electric Arc Furnace Using Continuous Scrap Pre-Heating System

BACKGROUND

An electric arc furnace (EAF) involves a high-temperature melting operation with melt inside the furnace and waste gases/off-gases from the furnace, operating at an average temperature of 1,650°C and 900–1200°C, respectively. The energy carried away by the off-gases is around 20% of the input energy which is alarming. This energy can be recovered by reusing as pre-heating the scrap. Following are the advantages of EAFs.

- Helps remove moisture from gas.
- Reduces electrode consumption.
- Reduces refractory consumption.

There are generally the following two types of scrap pre-heating technologies.

- Bucket pre-heating system.
- Continuous scrap pre-heating system.

CONTINUOUS SCRAP PRE-HEATING SYSTEM

The energy saving here is around 12%. The scrap is put on the conveyor and is passed through the pre-heating section, where the off-gases from the EAF are routed through the pre-heater in counter-flow direction. The additional advantages are as follows:

- Reduced harmonic and flickers
- Reduction in dust generation

Figure 5.5 shows continuous scrap pre-heating.

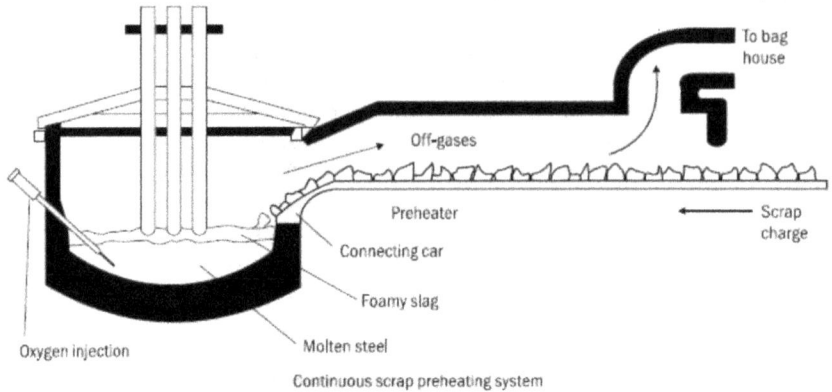

Continuous scrap preheating system

Source: http://infohouse.p2ric.org/ref/10/09048.pdf

FIGURE 5.5   Scrap pre-heating of an electric arc furnace using a continuous scrap pre-heating system.

TECHNO-ECONOMIC EVALUATION

Temperature of off-gases	800°Celsius
Temperature gain by the scrap	300°Celsius
Present specific energy consumption (SEC)	725 kWh/t
SEC reduction with pre-heating	80 kWh/t
Energy saved per heating cycle	522 kWh
Energy saving/day @ eight chargings/day	4176 kWh
Annual energy saving	1,461,600 kWh
Annual running days	350
Cost saving/annum	1.40 crore INR
Investment required	1.20 crore INR
Simple payback period	Ten months
Return on investment	117 %

Case Study 5.4: Bottom Stirring and Inert Gas Purging in an Electric Arc Furnace

BACKGROUND

The molten metal in an arc furnace may not be of homogenous mass or uniform quality across the cross-section, thus increasing the tap-to-tap (TTT) time and energy consumption, and potentially also leading to high rejection levels. The bottom stirring of the liquid bath in an EAF is a potential solution for better homogeneous mass and ensures uniform quality, which is accomplished using inert gases such as argon or nitrogen. The bottom stirring system based on inert gas injection is available either as a single tube or multi-hole plugs. These plugs are either buried in the furnace hearth ramming mix, or "indirect purging"; or in contact with still melt, or "direct purging." An indirect purging arrangement offers an improved stirring arrangement due to better distribution of inert gases. The bottom stirring using inert gases is more suitable for smaller furnaces. Bottom stirring further accelerates chemical reactions between steel and slag. The stirring helps in an increased heat transfer with an estimated energy saving of 3%. It further leads to increased metal yield of about 0.5%. However, the use of inert gas would require significant maintenance after every heat. The advantages are as follows:

- Improved control of the temperature and chemical composition
- Lower consumption of refractory and electrode
- Shorter TTT times
- Improvement in liquid metal yield

This image shows a bottom purging system.

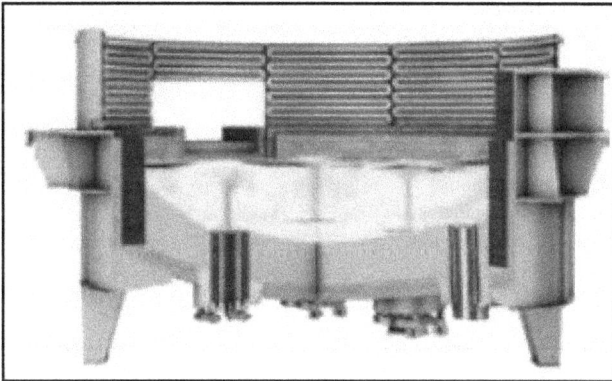

PHOTO 5.3 Bottom purging system.

Diagrams of direct and indirect contact stirring systems are given in Figure 5.6.

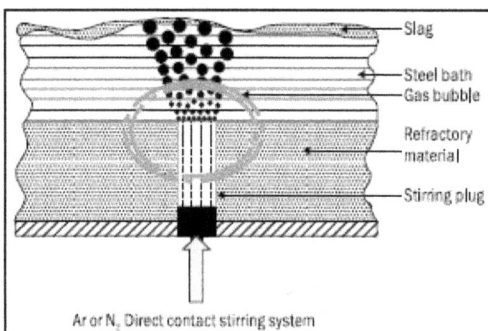

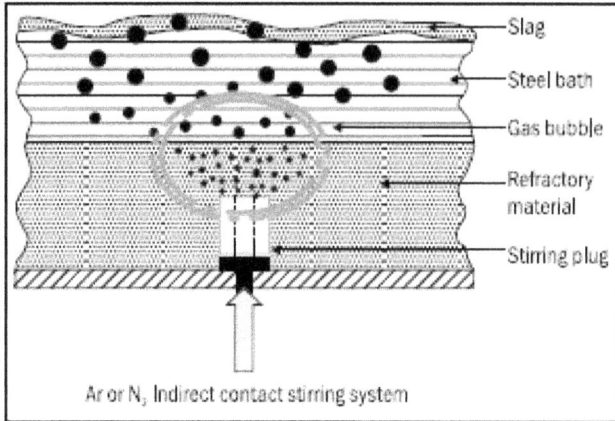

FIGURE 5.6   Diagrams of direct and indirect contact stirring systems.

TECHNO-ECONOMIC EVALUATION

Estimated reduction in SEC	21 kWh/t
Avg. liquid metal yield per heat	Five tons
Energy saving estimated per heat	100 kWh
Energy savings/day @ eight chargings/day	800 kWh
Annual energy saving	1461600 kWh
Annual running days	350
Cost saving/annum	26.88 lakh INR
Investment required	10.00 lakh INR
Simple payback period	Four months
Return on investment	268.8 %

Case Study 5.5: Use of Coherent Jet in Oxygen Lancing

BACKGROUND

In EAF, the oxygen injection system increases the use of chemical energy to enhance productivity. This injection is in the form of supersonic jet, which has the following two disadvantages.

- Formation of cavity in liquid metal bath.
- It creates splash, which leads to higher refractory consumption.

These shortcomings are eliminated by coherent jet, consisting of four co-jets mounted on furnace side walls. It acts as multi-purpose systems which acts as burner, lance and post-combustion device. In co-jets, compact lancing and decar-burizing can be achieved by keeping the oxygen stream coherent, i.e., retaining its original diameter and velocity over longer distances. The advantages are as follows:

- Better penetration of liquid metal bath (80% more than traditional lance)
- Supplies precise amounts of oxygen to liquid metal baths
- Less splash and cavity formation
- Lower refractory consumption
- Improves slag foaming with less carbon
- Decrease in air infiltration, as the slag door will be kept closed
- Reduction in average energy consumption to the tune of 5%
- Reduction in electrode consumption to the tune of 7%
- Reduction in TTT time by 10%

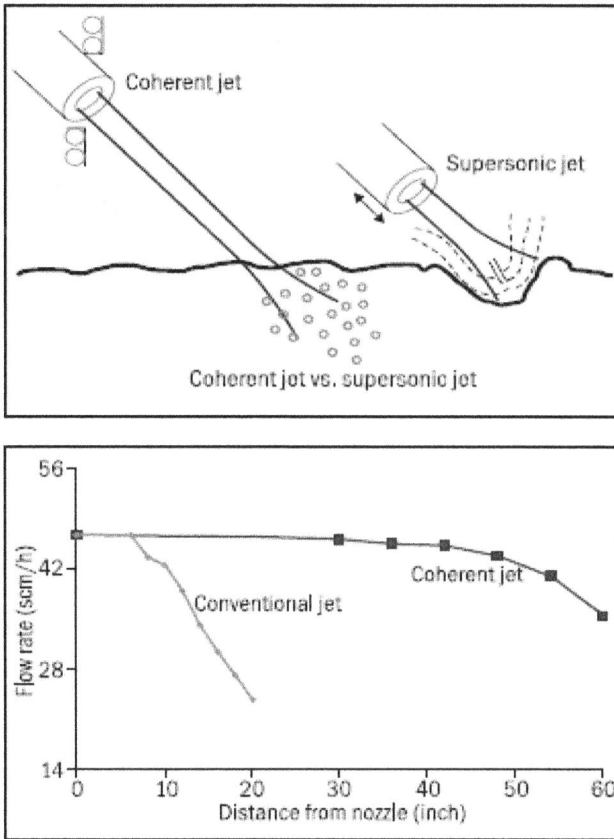

FIGURE 5.7   A comparison of coherent jet with conventional supersonic jet.

TECHNO-ECONOMIC EVALUATION

Estimated reduction in energy consumption	5%
Avg. energy consumption per heat	4,000 kWh
Energy saving estimated per heat	200 kWh
Energy savings/day @ eight chargings/day	1,600 kWh
Annual energy saving	560,000 kWh
Annual running days	350
Cost saving/annum	53.76 lakh INR
Investment required	1.20 crore INR
Simple payback period	27 months
Return on investment	44.8 %

Case Study 5.6: Improving Refractory of the EAF and Saving Electrical Energy by Avoiding Convection and Radiation Loss through Furnace Skin

BACKGROUND

The consumption rate of refractories in an EAF is about 20–35 kg per ton of liquid metal for smaller furnaces without water-cooled roofs and sidewalls. For furnaces with water-cooling systems, the refractory consumption is considerably low, about 5–10 kg per ton of liquid metal. High refractory consumption in an EAF leads to high downtime, resulting in lower productivity and higher production costs.

The use of improved refractories, such as castable alumina and mag-carb refractories, helps in improving the overall life of a furnace; for example, the life of sidewalls with the use of mag-carb refractory is about 200 heats without any repairs in the case of a continuously operated furnace. This reduces the downtime and improve productivity. The average energy saving from one EAF by improving refractory is calculated in what follows:

Heat Loss Through Ceiling

Parameter	Unit	Actual	Expected
Ambient temperature	°C	35	35
Average ceiling temp	°C	170	55
Convention factor for ceiling		2.8	2.8
Emissivity		0.5	0.5
Convection loss	kCal/m²/hr	1,288.47	118.43
Radiation loss	kCal/m²/hr	720.15	62.83
Total heat loss through ceiling	kCal/m²/hr	2,008.62	181.26
Actual heat loss	kCal/m²/hr	1827.36	

Ceiling area	m²	4.68	
Actual heat loss through ceiling	kCal/hr	8,552.0448	
Actual electrical energy loss	kWh/hr	9.94	
Heat Loss Through Body Wall			
Parameter	**Unit**	**Actual**	**Expected**
Ambient temperature	°C	35	35
Average body wall temp	°C	138	55
Convention factor for body wall		2.2	2.2
Emissivity		0.5	0.5
Convection loss	kCal/m²/hr	721.89	93.05
Radiation loss	kCal/m²/hr	476.66	62.83
Total heat loss through body wall	kCal/m²/hr	1,198.55	155.88
Actual heat loss	kCal/m²/hr	1,042.67	
Body area	m²	4.68	
Actual heat loss through body	kCal/hr	4,879.6956	
Actual electrical energy loss	kWh/hr	5.67	
Total electrical energy loss per heat	kWh/hr	15.61	
Average heat time	hr	4	
Average electrical energy loss per heat	kWh	62.44	

TECHNO-ECONOMIC EVALUATION

Energy saving from one EAF per heat	62.44 kWh
Energy savings/day @ eight chargings/day for two EAFs	499.5 kWh
Annual running days	350
Annual energy saving	174,925 kWh
Cost saving/annum	16.78 lakh INR
Investment required	15.00 lakh INR
Simple payback period	11 months
Return on investment	112 %

Case Study 5.7: Improving Refractory of the 10 T Heat Treatment Furnace and Saving LDO by Avoiding Convection and Radiation Loss through Furnace Skin

BACKGROUND

In Tata Steel Limited (TSL), three heat treatment furnaces operate at 10 T, 12 T and 20 T capacity, out of which the refractory material of 10 T HTF has been deteriorated and significant amount of can be saved by improving the refractory material. The amount of saving potential is given in what follows.

Heat Loss Through Side Wall of 10 T HTF

Parameter	Unit	Actual	Expected
Ambient Temperature	°C	35	35
Average side wall temp	°C	180	55
Convention factor for side wall		2.2	2.2
Emissivity		0.5	0.5
convection loss	kCal/m²/hr	1106.96	93.05
Radiation Loss	kCal/m²/hr	807.92	62.83
Total heat Loss through side wall	kCal/m²/hr	1914.88	155.88
Actual Heat loss	kCal/m²/hr	1759	
Wall Area (4 side)	m²	63	
Actual Heat Loss Through side wall	kCal/hr	110817	
Actual LDO consumption	Liter/hr	11	
LDO saving potential	Liter/hr	11	
Average heat time	hr	25	
Average LDO loss per heat	Liter	275	

TECHNO-ECONOMIC EVALUATION

LDO saving per heat	275 liters
LDO cost saving per heat	22,000 INR
Energy saving/day @ 150 heats per annum	33.00 lakh INR
Investment required	18.00 lakh INR
Simple payback period	7 months
Return on investment	183%

Case Study 5.8: Adopting VFD in Overhead Cranes

In VISL material handling, i.e. movement of material from one place to another within the plant, is required. Overhead cranes are used for this purpose. Depending upon the weight of the material to be handled, various capacity cranes are used in all the departments. The capacity of the crane is indicated in tons i.e., the max load which can lift. In VISL cranes of 2–50 tons are used. The details of crane are as follows:

Capacity: 40 T
Crane no: 321
Used for handling hot steel

TABLE 5.2

Motor Details

	Slip Ring I.M. Capacity in KW	Rotor Voltage in Volts	Rotor Current in A	Rotor Resistance in ohms	Speed in RPMs
MH	50	200	156	0.77	725
AH	24	220	85	1.50	965
Trolley	10.5	190	36	3	965
Bridge	32	300	67	2.60	970

Duty cycle is 60% and 24 hours working per day.

TABLE 5.3

Energy Analysis

Average power dissipated in KW	MH	AH	Bridge	Trolley
$3 \times I \times I \times R$	56.21	32.51	35.01	11.66
Total load in KW	135.4			
Energy dissipated in external resistances per day, considering 60% duty cycle	$(0.6 \times 24 \times 135.406) = 1949.8\ \text{kWh}$			
Annual wastage of energy	7,11,693.93 kWh			
Cost of energy at 6 Rs per unit	Rs 42,70,163			
Total annual saving including spare charge in Rs	Rs 46,00,000			

Source: Nandini (2016)

From this analysis, it is found that approximately 700,000 kWh energy is wasted in the external resistance connected to the rotor winding.

Case Study 5.9: Waste Heat Recovery from Forging Furnaces Using Air Pre-Heaters

BACKGROUND

Forging furnaces are in operation to pre-heat steel blocks up to 1,100°C before drop forging operations. The furnace exhaust temperature is around 500°C.

PRESENT SCENARIO

A WHR system is being installed to pre-heat the combustion air for the furnace by recovering some of its waste heat in the exhaust air. The fresh air for

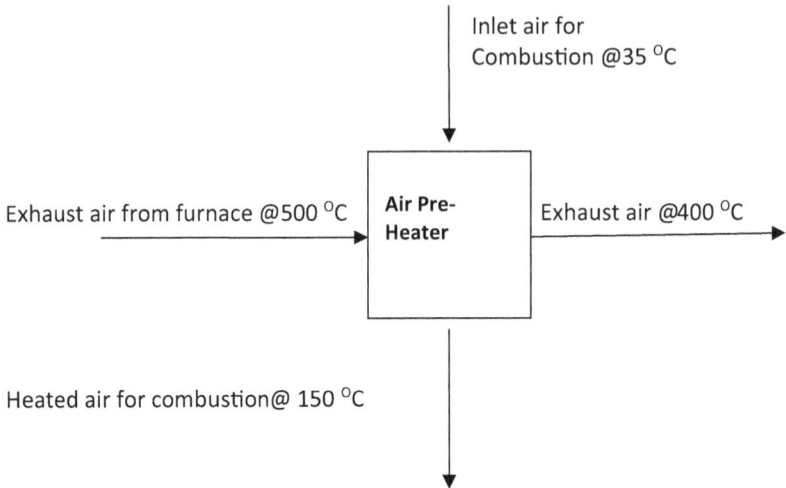

FIGURE 5.8   Combustion air pre-heater from flue gases.

combustion is being pre-heated by means of circulating the same through an air-to-air heat exchanger. Every 20°C rise in temperature of fresh air will improve the efficiency of the furnace by 1%. Therefore, it is estimated that around 5% savings in fuel is being achieved by installing combustion air pre-heaters from flue gases.

The flow diagram depicting this is shown in Figure 5.8.

SUMMARY

Based on the average production rate of 1250 T/month and specific gas consumption of about 100 SCM/T @ Rs. 29/SCM, it is estimated that annual savings being achieved by the plant due to installation of a waster heat recovery system to pre-heat combustion air is around Rs. 18.0 lakhs/year (considering 5% savings in gas consumption).

Source: CII (2022).

5.14   CONCLUSION

An energy audit is the key tool to assess and identify the EE&C potential with techno-economic viability of each recommendation of energy conservation opportunity or energy-saving opportunity in a metal industry. The energy audit types, procedures, needs and implementations outlined in this chapter would be handy reference to practitioners and researchers in this field. The discussions have been supported with nine case studies of major areas of higher potential for EE&C in metal industries. Related documents on manufacturing processes of metal industries are presented in Chapter 4, where these energy audit toll and techniques are being aptly applied.

REFERENCES

ACEEE (2021); Summer Study on Energy Efficiency in Buildings, American Council for an Energy Efficient Economy (ACEEE), www.aceee.org/files/proceedings/2016/data/papers/12_37.pdf, accessed in 2021.

Avina John M. and Rottmayer Steve P. (2016); Abraxas Energy Consulting, ©2016, Virtual Audits: The Promise and the Reality.

BEE (2016); Small Industries Development Bank of India (SIDBI) New Delhi – 1st Prize Winner Among Financial Institutions; Bureau of Energy Efficiency (BEE), http://knowledgeplatform.in/wp-content/uploads/2016/01/Small-Ind.-Dev.-Bank-of-India_1st-Prize.pdf

BEE (2018); Enhancing Energy Efficiency through Industry Partnership PAT, Bureau of Energy Efficiency (BEE), https://beeindia.gov.in/sites/default/files/press_releases/Consolidated%20Report.pdf

BEE (2022); The Energy Conservation Act, 2001 (Amended in 2010), Bureau of Energy Efficiency (BEE), https://beeindia.gov.in/sites/default/files/The%20Energy%20Conservation%20Act%2Cchp1.pdf, accessed in 2022.

CII (2022); Best Practices – Energy Efficiency in Forging Industry – document prepared for "Financing Energy Efficiency at MSMEs" project; CII – AVANTHA Centre for Competitiveness for SMEs, Confederation of Indian Industry (CII); https://fdocuments.in/download/best-practices-energy-efficiency-in-forging-best-practices-energy-efficiency accessed in 2022.

FICCI (2018); Indian Non-Ferrous Metals Industry – Way Forward, Feb-2018, FICCI; https://ficci.in/spdocument/22947/Non-Metal-Report.pdf

IBEF (2009); Metals Industry in India, India Brand Equity Foundation (IBEF), https://www.ibef.org/download/Metal_171109.pdf

Jarza S (2011); Importance of Energy Management in Foundries. *Polish Journal of Management Studies*, Vol. 4; https://bibliotekanauki.pl/articles/406161

Nandini K.K. (2016); Energy Audit in Vishveshwarayya Iron & Steel Plant, NITTE Conference on Advances in Electrical Engineering NCAEE-2016; https://ijireeice.com/special-issue-ncaee-16/

Teri (2015); Enabling Finance for Scaling up Energy Efficiency in MSMEs TERI, http://cbs.teriin.org/pdf/Energy_Efficiency_Final.pdf

6 Implementation of Energy Efficiency and Conservation Projects

Jitendra Saxena

6.1 INTRODUCTION

Energy efficiency and conservation projects, and support of energy efficiency companies, are very significant for sustainable development to all global commodities with high degrees of reliability, affordability and achieving the mission of energy to all (IEA, "World Energy Outlook 2015,"). The energy demand will be reduced by the pivotal role of energy service companies (ESCOs) who will provide the client all technical and financial support to improve energy performance and energy efficiency. Every country is applying the traditional approach of ESCOs, and there is a heterogeneous model for ESCOs for various countries depending on their environment, culture, technology and availability of resources (Hansen and Associates, 2011).

This chapter will provide an in-depth study of ESCOs which are operational in different developed and underdeveloped countries and alliances like the United States, the European Union, China, India and Japan. It is of paramount importance to study different techniques of ESCO used for energy efficiency promotions, its present conditions and its progress during the energy project cycle, and its judicious applicability in developed and developing countries. The organizational-based management systems, accounting management and the reducing energy losses are focus areas and objectives of energy-saving projects in metallurgical industries. The success stories for energy efficiency and conservation projects highlight efficiency, operating costs, losses, etc. It is difficult to estimate resources of energy holistically and to calculate the energy intensity of a plant. The basic fundamentals are to create methodological studies for project management and energy efficiency, and conservation programs on metallurgical industries, focusing on the management of resources. The chapter provides fundamental studies of operational procedures, planning and energy management to managing projects for metallurgical systems. The supporting energy-saving factors should be well structured to calculate the efficiency of the metallurgical enterprise and will affect the process of the energy program. A detailed program of energy efficiency and conservation for a metallurgical industry should cover such areas as developing energy baseline, energy accounting system, energy use by consumers and its improvement, reducing transmission, energy consumption monitoring, technological processes and units innovation and technological advancement, energy

DOI: 10.1201/9781003157137-6

balance improvement of the industries, and staff involvement in energy-saving and conservation activities.

6.2 ENERGY SERVICE COMPANIES IN METALLURGICAL INDUSTRIES

Energy management programs were framed with fundamental procedures to manage energy consumption of metallurgical plants, with sustainable development of energy resources in metal industries. They can provide improvement in the efficient utilization of resources (internal and external), cost control and product quality, and the overall competitiveness of metallurgical industries. Project management knowledge systems are very good inspiration for the metallurgy industry in all aspects (IEA, "World Energy Outlook 2015,"). Project management's overall objective is to satisfy the requirements and demands of customers once the project is in operation—the control, satisfying client requirements and executing all activity of the project efficiently. The key points of effective controls are to measure, supervise and monitor the progress of the plan, periodically and correctly. The modern project management method can resolve problems such as low efficiency of the steel industry and the implementation of effective control plan. An ESCO is an energy services company which provides energy project design, implementation of energy efficiency projects and retrofits after identification of energy-saving potential through energy audits with the present infrastructure. Energy-saving solutions to clients and customers are provided by ESCOs. This includes wide spectrum of services, including the following.

- Outsourcing of energy.
- Energy management, efficiency and conservation.
- Risk management.

Due to increases in oil prices in the early 1970s, the concept of energy efficiency and conservation projects with ESCOs come into force—being cost-effective and optimum use of energy of paramount importance. The main function of Energy Service Company is to provide service to customers by enabling availability of energy at minimum cost, related management, financing, performance monitoring and contracting. The most important business models for ESCOs are guaranteed saving and shared savings models which are adopted across the globe.

The characteristics of ESCOs are as follows.

- Minimizing utility costs.
- Assurance of energy-saving effects by ESCO.
- Provision of complete services.
- Comprehensive verification of energy-saving effects.
- Financing environments not based on assets.

6.3 OBJECTIVE OF ESCOs

ESCO projects with an investment payback period of 0–2 years can be executed, and these cream-skimming projects are taken as a priority in order to achieve short-term

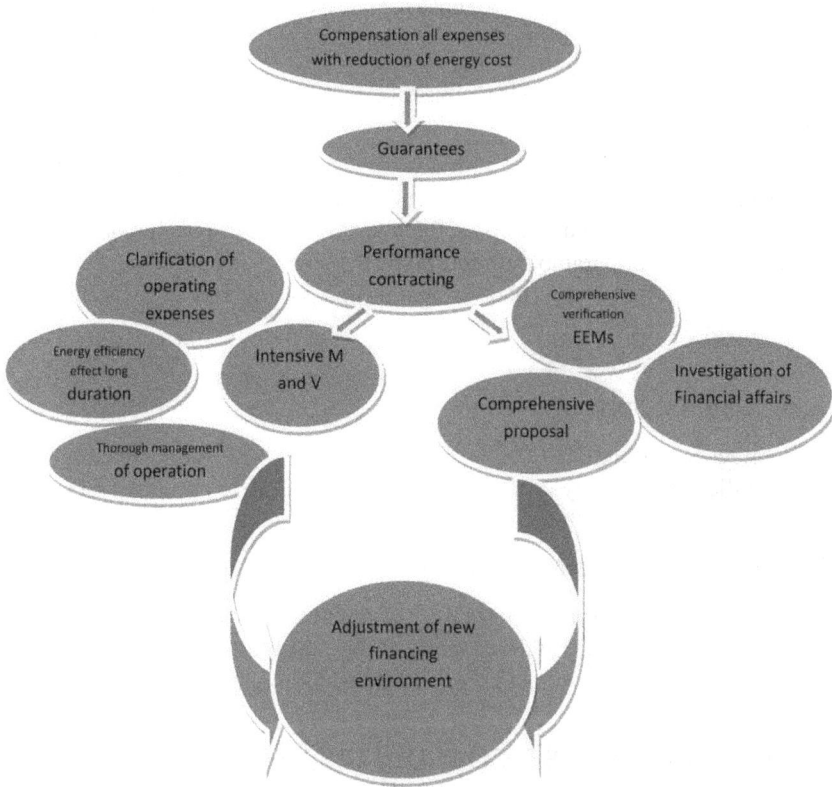

FIGURE 6.1   Key features of implementation of energy performance contracts.

energy goals. The target of ESCOs also may be the energy projects with an payback period of 2–3 years. Energy efficiency renovations with a payback period of less than two years should be considered whose role is to contribute to the global economy (Hansen, 2003). A matured ESCO market will be achieved only after this concept is shared by the financial institutions and the demand/supply side of the service. Provision of incentives, such as tax incentives, subsidies and low-interest loans, will also contribute to the expansion of the market. On the other hand, underdevelopment of the financial environment becomes a major constraint on the growth of the market.

6.3.1   ENERGY PERFORMANCE CONTRACTS (EPCs)

Figure 6.1 shows overall key features of performance contracting with guarantee including intensive measurement and verification (M&V) with operational and financial proposals. ESCOs had not been motivated as per the Indian perspective, but due to exponential demand of electricity and comfort levels, they are now growing in India, as well. If we want to optimize the electricity consumption of the commercial complex, utilizing the service of ESCOs can be undertaken to optimize electricity consumption and to facilitate reduction of expenditures on electricity (E. Commission, 2016).

6.3.2 Energy Saving

The ESCO's overall objective and vision is saving of energy, and it should be managed properly and effectively. During the energy audits process and implementation, the ESCO provides detailed reports of the consumption and installation of energy-saving devices. It is of paramount importance that energy is conserved significantly after execution of the energy plan recommended by the ESCO (Murakoshi and Nakagami, 2009).

6.3.3 Technical Support

If the contract is with an ESCO, all expenses incurred will be undertaken by ESCO for the installation of energy-efficient equipment; this is a vital point for EPCs.

6.3.4 Monitoring and Maintenance

Monitoring and maintenance responsibility also will be undertaken by the ESCO for equipment which is installed at the property, if the clause and services are mandatory and are mentioned in the contract between the ESCO and the customer.

6.4 PROGRESS OF THE ESCO INDUSTRY IN INDIA

The energy efficiency market in India is projected to be of INR 150,000 crore (i.e. almost USD 20 B), out of which only 5% potential has been utilize by ESCOs so far. The Bureau of Energy Efficiency (BEE) organizes facility programs for end-users, technology providers, facilitators, ESCOs, financial institutions and government agencies—together in a single platform—to accelerate energy performance project through ESCO route. BEE of India organizes different workshops on different case studies best practices for financing instruments for creating awareness between end-users on energy services available through ESCOs.

6.4.1 Energy Conservation by Energy Source in India

Pushed by the oil crisis in the 1970s, the Petroleum Conservation Research Association (PCRA) was established under the Ministry of Petroleum and Natural Gas in 1976, and it promoted energy efficiency in oil, as well as energy-efficient technologies of petroleum products, which was the beginning of institutionalized energy conservation efforts in India. The PCRA is now undertaking energy conservation activities not only in oil, but also in other types of energies. It is also expanding ESCO-related activities, such as ESCO projects, energy efficiency auditing, and ESCO seminars. In the meantime, the Department of Power, Ministry of Energy, set up the Energy Management Center (EMC) in 1989 and has planned and implemented power-related energy efficiency programs. From the late 1980s to the mid-1990s, in an effort to cover both oil and power disregard of energy source difference, energy conservation services were advanced by the industrial associations like the Confederation of Indian Industry (CII) and the National Productivity Council (NPC), in cooperation with research organizations like the Energy and Resources Institute (TERI), Central Power Research Institute (CPRI), etc.

6.4.2 START OF FINANCING TO ENERGY EFFICIENCY PROJECTS BY FINANCIAL INSTITUTIONS

The Indian Renewable Energy Development Agency Ltd. (IREDA) was established in 1987 as the governmental financial institution aimed at promoting dissemination of and financing to the technologies and projects associated with new energy and energy efficiency. IREDA has since provided financing to the introduction of new energies, implementation of energy efficiency measures and formulation of ESCO finance, with the financial assistance from foreign governments (Dutch government: 18 million guilders; Asian Development Bank [ADB]: USD100 million; World Bank [WB][the first time] (Okay and Akman, 2010; World Bank, 2016): USD145 million; Denmark: USD15 million; Germany: 6,135 Euros; WB [the second time]: USD135 million. In the early days after its establishment, focus was placed on new energy-related projects, but energy efficiency projects (including ESCO projects) began to be implemented around six years ago. The total number of such projects reached 17 and its total loans amounted to 1.71 billion rupees (Rs) (4.4 billion yen). Also, with financial assistance from the Global Environment Facilities-Technical Assistance Program (GEF-TAP) (subsidy: USD5 million), it also promoted activities such as establishment of IREDA's institutional systems, provision of technical assistance, energy conservation marketing, improvement of DSM investments and encouragement of entry of the private sector into the energy efficiency projects. As part of these efforts, other activities were also taken which include creation and expansion of the ESCO market; capacity building of ESCO-related enterprisers; implementation of energy efficiency projects in such industries as steel, pulp and paper, textiles and hotels; and preparation of energy conservation investment manuals for energy-intensive companies and small and medium-sized companies.

In addition, other development-related financial institutions, such as the Industrial Development Bank of India (IDBI), Small Industries Development Bank of India (SIDBI) and the Industrial Credit and Investment Corporation of India (ICICI), began to lend to energy efficiency projects from around the mid-1990s, with its own funds or with financial assistance from international organizations (Taylor, et al., 2008).

6.4.3 START OF ESCO PROJECTS

India's first ESCO-related program was the feasibility study (FS) of an ESCO project which was undertaken in 1992–1993 with financial assistance from the United States Agency for International Development (USAID). Then, in 1994–1995, an

ESCO business promotion program was implemented with the help from the Energy Management Consultation and Training (EMCAT) Program of the USAID, using a subsidy provided from the US agency to the IDBI. In 1995–1996, the USAID organized two ESCO inspection/observation visits, an Indian group to the United States and a US group to India. Through these supports by the USAID, several ESCO providers came into existence in India. In 1999, a four-year Energy Conservation and Commercialization (ECC) project of the USAID started. The primary purpose of the ECC project is to promote creation of the market for energy-efficient technologies and energy conservation services in India. It is a project to provide technical assistance and training to the government for the formulation of highly market-oriented policies, and to help capacity building of utility companies and private companies for the introduction of energy-efficient technologies and implementation of DSM, while removing institutional and technical barriers to the dissemination of energy conservation, so that highly market-oriented energy efficiency business emerges in India eventually. In the first phase (ESCO-I: 1999–2000), several energy efficiency projects, ESCO projects, and DSM projects were implemented (Ellis, 2009). In the second phase, in addition to these projects, activities such as the following are being implemented: support to the BEE, DSM projects at utility companies and preparation of energy conservation building codes (the entire nation is divided into six zones). The organizations associated with the ESCO project are consulting firms, ICICI and the Ministry of Power. BEE has published a list of 125 accredited ESCOs in India [1].

6.5 ESCO PROJECTS IN DEVELOPED AND DEVELOPING COUNTRIES

The trends which were used in past for ESCOs for operation and implementation of the process of installing energy-efficient equipment and financing methods in the world follow the ESCO model perused in the United States. Figure 6.2 shows holistically ESCO implementation plan followed globally in every country with its ten points important features (E. C.-J. R. Center, *European ESCO Market Survey*, 2012; LLC, 2013).

The ESCOs are executed in four ways: (i) funding resources and financial institutions are selected; (ii) stakeholders participation is secured; (iii) standards and reporting for energy policy regulation are developed; and (iv) energy efficiency performance business and targets with contract assignments is evaluated (CCAP, 2012; Paolo et al., 2006; Vine, 2005; Lee et al., 2003).

Japan followed ESCO standards and norms strictly and focused on industries, as they are key market players and driving forces for ESCO projects. Japan's funding mechanism is good for large companies but is less attractive financing for small and medium-scale companies. For ESCO and EPCs in Australia, the key players are municipal, commercial and industries. In Austria, there are many companies which promote energy efficiency in the field of space heating, control and automation, and building envelope projects.

Outline of ESCO

Global situation of ESCO companies /market

Successful model of ESCO type Energy
conservation promotion in developing countries

Successful model support system to ESCO type
Energy conservation promotion in developing

Evaluation of support approach to ESCO type Energy
conservation

Domestic and International resources

Conception of new support plan specifically related
to ESCO type

Point to note on cooperation in ESCO type Energy
conservation promotion

Proposal letter reflecting Thermatic

Proposal of tie up with the field of global
warming counter measures.

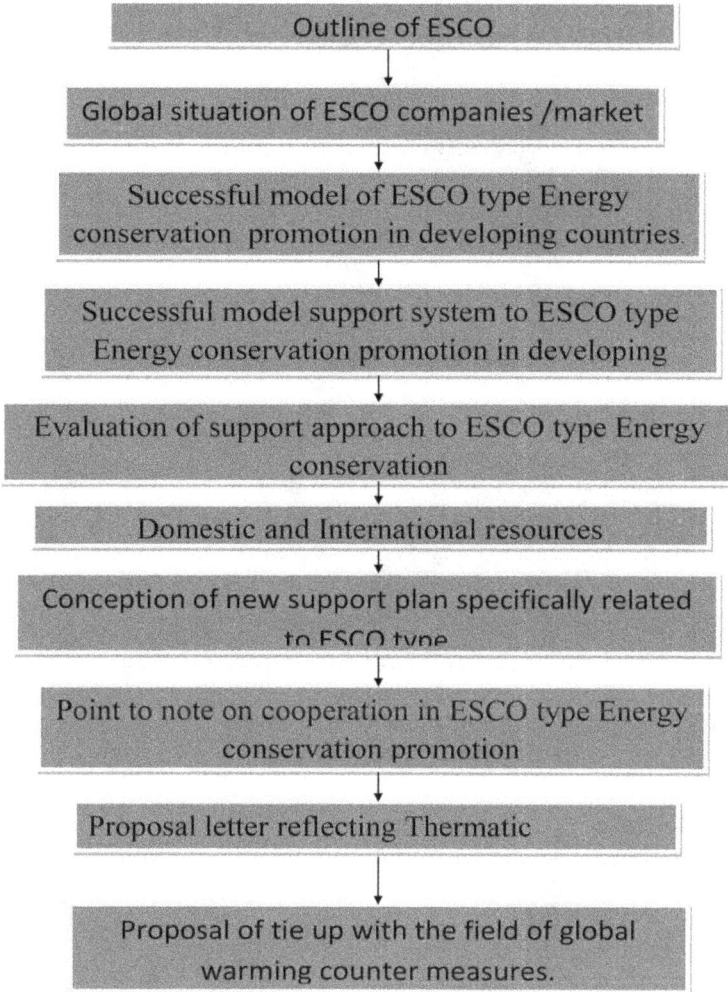

FIGURE 6.2   Energy service company research projects implementation plan flow chart.

There are financial, institutional and technology barriers among several complications to ESCO implementation in China (Taylor et al., 2008 and Da-li G., 2009).

Thailand has attractive financial incentives due to successful ESCO implementation, and drivers are from small and medium-scale businesses in the building and commercial sectors. Table 6.1 and Table 6.2 summarize the ESCO projects in both developed and developing countries (Murakoshi and Nakagami, 2009). In Africa (Egypt, Kenya and South Africa), conducting cooperation with international institutions for creating funding, increasing public awareness and creating ESCO associations are strategies for actualizing ESCO in their countries. In Brazil, the development of an ESCO association and cooperation with the Canadian government enhanced

TABLE 6.1
Energy Service Company Projects for Developed Countries

Implementation

Types of Project	Regulatory Factors	Drivers for Promoting ESCO	Financing
The long payback period for many ESCO projects Large-scale ESCOs and customers	Strengthening standard and regulation Starting in public facilities	Private sector facilities (industrial sector) Japanese Association of ESCOs (JAESCO)	Shared savings contracts (90%) Incentives (e.g. subsidies) Excellent financing mechanisms for large-scale customers
EPC in commercial, municipal and industrial sectors	Government accreditation Development of EPC facilitators and guidance frameworks Standard contracts	Industry, commercial and municipal Australasian Energy Performance Contracting Association Limited (AEPCA)	Multiyear budget (five years) Treasury funds for repaying the guaranteed saving loans Commercial leasing arrangements
Space heating, AC, control and automation, lighting Public sector buildings	Certification and accreditation schemes Standard and ecolabels Mandatory of energy consultation Standardization contracts	Small and medium enterprises (SMEs) Dachverband Energie contracting Austria (Professional Association for Energy Contracting)	EPC Shared saving model Commercial banks financing
Public ESCO, Fedesco Public sector buildings and private industry facilities projects Main target healthcare facilities, educational and office buildings	Establishment and funding of public ESCOs Information campaigns The obligation for recruiting energy managers and mobility officers	Four public ESCOs and 10–15 private firms (six large, 5–7 SMEs) Public ESCOs that act as market facilitators Slow growth BELESCO: the Belgian Association of ESCOs and energy service providers and AGORIA Green Building platform	Public energy service contracts Third party financing (TPF) EPC and Smart EPC (energy, maintenance, comfort and building value performance contracts)

All energy efficiency sectors	National and European legislation (key movers) The European directives	Numerous associations Around 550 companies so far involved with ESCO EPC market Increasing energy prices	80–85% energy supply contracting (ESC), 8–10% EPC Mainly shared savings model
The industrial sector, public buildings, hospitals, schools, offices, social housing **Supply side and networks (district heating), HVAC, control technologies, lighting, and public lighting**	Climate and energy conservation policy Financial incentives	Various trade associations Cost and environmental motivations	The Green Investment Bank. EPC (both shared savings and guaranteed savings)

ESCO development. The ESCO industry in the developed countries and alliances like the United States, European Union, France and Germany has been largely affected by peak oil crises and environmental conditions (Marino, 2010 and Morgado, 2014).

The characteristics observed in the recent swings are expansion of the federal market, development of new services based on information technology and the return to the energy conservation business. ESCO business in the United States started as a business model which appeared due to the first oil crisis in 1973 and the second in 1979.

ESCO providers in the initial phase were engineering consultants who offered energy audit services. They proposed improved energy systems and improved energy management based on the details of the energy audit results. However, this was not established as a business. The origin of ESCO is the providers which developed mere consulting services into project development services. In the 1980s, building management companies and control system manufacturers set up their energy service divisions that offered performance contracts and participated in ESCO business. In the same period, ESCO providers who took part in utility DSM programs started to be involved in the United States and other developed countries (Marino, 2010 and Morgado, 2014).

The traditional ESCO operations in the world follow the ESCO concept used in the United States, which includes design engineering, energy-efficient equipment installation and financial methods. ESCO operation is quite identical in all aspects, as per led-down procedures and operations of ESCO compared to fundamental ideas of ESCO in the United States (Bertoldi, 2014).

Japan followed standards and regulations for developing ESCOs strictly and focused on industries as they are driving forces for ESCO projects and key market drivers. Japan is the best provider of funding mechanisms for big companies but lesser financing mechanism for small and medium-scale companies. For ESCO and EPCs in Australia, the key players are industries, commercial enterprises and municipalities (Paolo et al., 2006). There are 50 ESCOs which are working for energy service companies for the promotion of energy efficiency in the fields of space heating, air conditioning, control and automation and building envelope projects.

The ESCO market in the United Kingdom and most of the ESCOs were responsible for financing, installing, operating and maintaining PV systems rather than energy efficiency. ESCOs and EPC-specific regulation do not pertain in the UK, and the business model is dynamic in the UK (Paolo et al., 2006).

6.6 ACTIONS OF THE INDUSTRIAL SECTOR TO ESCO BUSINESS

The Resource Conservation & Management Group which was set up under the Federation of Indian Chamber of Commerce (FICCI) for promoting efficiency improvement of industrial processes, effective utilization of resources, waste management, etc., has consolidated energy conservation activities voluntarily, impelled by the enactment of the Energy Conservation Act in 2001. Specific

TABLE 6.2

Energy-Saving Conservation Projects for Developing Countries

Country and Source/Remarks	Implementation Types of Project	Regulatory Factors	Drivers of Promoting ESCO	Financing
Korea	Street lighting, industry	Street lighting as a pilot project Multiyear procurement	High energy prices Korean Association of ESCOs (KAESCO)	Tax credits and long-term and low-interest financing
China G. Da-li, "Energy service companies to improve energy efficiency in China: Barriers and removal measures," *Procedia Earth and Planetary Science*, pp. 1695–1704, 2009.	Green light, heating network, boiler retrofit projects, central air conditioning, integrated building and industrial sectors	Founding Energy Conservation Information and Dissemination Centre (NECIDC) Creating ESCO Association and ESCO partial loan guarantees Revision of the Energy Conservation Law	Projects for reducing emissions China Energy Conservation Service Industry Association (EMCA)	EPC (shared savings, guaranteed savings, and outsourcing management contract). Short payback period (less than two years) Loan guarantee system
Thailand	Industrial, commercial sector	Providing incentives for ESCO Conducting a pilot project for four industrial facilities	Electricity Generating Authority of Thailand (EGAT) as a role of ESCO promotion	EPC (shared and guarantee savings) Energy Conservation Promotion Funds (ENCOND Funds) Revolving fund mechanism with the low-interest rate (less than 4%)

(Continued)

TABLE 6.2 Continued

Country and Source/Remarks	Implementation Types of Project	Regulatory Factors	Drivers of Promoting ESCO	Financing
India	Industrial, commercial sector	Small project ESCO for municipal retrofit lighting Improve energy efficiency Involving international organizations ESCO promotion	Building and industry sector The Federation of Indian Chambers of Commerce and Industry (FICCI) The Indian Council for Promotion of Energy Efficiency Business (ICPEEB)	EPC Decentralized investment of ESCOs Low-interest financing
The Philippines	Lighting retrofitting	Developing a model ESCO contract	Lighting (street lighting). The IFC/GEF Efficient Lighting Initiative (ELI) program Education for retail banks in EE project financing	EPC Development Bank of the Philippines (DBP)
Africa (Egypt, Kenya, South Africa)	Industrial, commercial (Egypt) Industrial (Kenya) Residential, industrial, commercial, agricultural, municipals (South Africa)	In Egypt, providing training for energy audits, energy efficiency technologies, and ESCO projects evaluation and financing In Kenya, conducting a survey banks In Kenya, hired energy consultant to explore establishment of ESCOs	Credit guarantee (Egypt: joint cooperation with US Agency for International Development) Egyptian Energy Service Business Association (EESBA) In Kenya, creating Kenya Association of Manufacturers Industrial Energy Efficiency Project In South Africa, creating ESCO Association (the South Africa Association of ESCOs [SAAEs], and the Black Energy Service Companies Association)	EPC Cooperation with an international organization to get funds/credit.
Brazil	Industrial, commercial, municipal, residential sector Greenhouse Gases Emissions Reduction in Brazilian Industry (GERBI) Project	Establishing National Electricity Conservation Program (PROCEL) to fund energy efficiency project Collaboration with Canadian government	Developing ESCO Association in 2009 (Brazilian Association of ESCOs [ABESCO])	Loan guarantee fund Creating a fund and joining cooperation with PROCEL

activities implemented include energy audits; holding of energy conservation seminars, workshops and meetings; and energy conservation trainings. In 2005, it held the first International ESCO conference and a workshop on M&V in New Delhi, in cooperation with PCRA. It also implemented several ESCO pilot projects. Soon after the first International ESCO Conference, major ESCO providers in India assembled and agreed on the establishment of an ESCO industry association to vitalize the ESCO industry. Deliberations continued, and in January 2006, the Indian Council for Promotion of Energy Efficiency Business (ICPEEB)—which is comprised of various entities related to ESCO industry, ranging from ESCO providers, BEE-certified energy auditors and energy managers, manufacturers and dealers of energy-efficient equipment, to financial institutions—was established. The founding members are 16 ESCO companies. Major activities include information exchanges among ESCO-related providers, seminars, awareness-raising activities, promotion of application of the International Performance Measurement and Verification Protocol (IPMVP) and technical assistance to relevant governmental organizations.

China and Korea based on the principle of optimum utilization of energy use by adopting energy security and energy use will be fully sustainable countries. Optimum utilization means it should be cheap, clean and easily affordable to the general population. Energy sustainability means that fossil fuels for the plenty amount of use will deplete within 100 years and therefore shift to nonconventional sectors. Therefore, world initiatives are there to transition from fossil fuels to nonconventional energy sectors. Environmental sustainability is another useful factor, and carbon footprints are the key driven technology and benchmark parameter for the economic growth. Renewable energy sources and the industry sector is the largest consumer of electrical energy, and the transport sector is also a foremost consumer of energy. The major consumer of energy in the industrial sector is the metal industry, within which copper and aluminum manufacturing operations are more significant consumer of energy.

Energy use and energy sustainability are the driving factors for optimum utilization and energy conservation measures for metal industry as mentioned below.

1. Energy use: Energy use in metal industry is very high of the order Gigajoules per Tons of production and energy conservation measures are of paramount importance and driving factor for profitability and sustainability of metal industry.
2. Energy sustainability: The process involved in the metal industry should be energy efficient and identified energy conservation measures are to be implemented for optimum utilization of energy.
3. Energy security: The IEA was created to ensure secure and affordable energy supplies, and it conducts analysis on current and future risks for oil supply disruption, emerging gas security challenges, and increasing system flexibility and resilience of the electricity sector. But energy transitions and the growth of cyber criminality have expanded the scope of what constitutes energy security.

The metal sector's carbon footprint and its effects are key driving factors of energy efficiency and consumption in the metal sector for example the aluminum, copper, Iron and steel industries. Energy audits are extensively used by designated consumers (DCs) and industries. Some of the countries' specific mandatory requirements are to conduct energy audits every two years with implementation of recommendations with PAT schemes, and this scheme is globally accepted. The energy projects' feasibility studies are done and identified projects are implemented if they are within the technical and financial acceptable limits—but some of the projects are not implemented, due to technical and financial limitations. Some projects cannot be implemented due to their emergency constraints or lack of innovation and risk-taking capability. Performance capacity with energy service company segments are exponentially increasing and becoming a key driven methodology in every part of the world.

There are different mechanisms available for frequency of EE and communication projects. Innovation and new initiatives act as a catalyst and provide financial gains and fulfill the long-term goals, while risks are reduced. M&V are used for performance evaluation of the project, which is explained in detail in Chapter 7.

Project implementation and execution is dependent on the following factors which are to be monitored for completion of project:

1. Feasibility study with risk analysis. Feasibility study of the project is the initial phase for starting a project which will identify all parameters which are related with project and how the risk which are identified are to be mitigated with robust supervision and monitoring
2. Analysis a technical and financial constraint. Technical and financial measures are to be identified and the constraints are to be investigated with proper solution
3. Project implementation and execution. It is very vital parameter for project identification, negotiation, appraisal, preparation and project execution.
4. Emergency evaluation and implementation. Evaluation and implementation of project involved process scheduling, task completion, sequencing and project completion.
5. Project performance and appraisal with measurement and verification.

After completing the energy audit report in all aspects, what will be our next course of action or what can we say should be the next course of action after submitting final energy audit report? The energy audit report provides a new project with investments and without investments. ESCO business in India has been advanced by a variety of organizations such as: governmental organizations like BEE, PCRA, etc.; banks like IREDA and the State Bank of India (SBI); ESCO providers; and FICCI. It presents one characteristic of India's ESCO business; namely, they are not evolving around some specific governmental line (Task in Asian countries, 2009).

The first international ESCO conference was held in New Delhi in June 2005, under the joint sponsorship of PCRA and FICCI (with the support from the Ministry of Petroleum and Natural Gas). Active discussions took place on such topics as successful ESCO business practices in foreign countries, barriers to ESCO business in India and measures for the promotion of the ESCO industry in the future.

6.6.1 FINANCIAL MODELS THROUGH WHICH ESCOs OPERATE

Financial models are needed by ESCO for effective decisions, and there are three parties involved in these models: financial agents (like investors, banks or creditors), the client and the ESCO.

There are basically two financial models which facilitate the functioning of ESCOs with clients (Figure 6.3 and Figure 6.4).

Figure 6.3 shows an energy sector management assistant program (ESMAP) based on the special purpose company model with special features of financing and equity.

In the model of guaranteed saving, the ESCO will confirm performance energy savings from the project, without financial responsibility, and therefore the financing to customers is done directly by the banks or financial bodies. In the "shared savings" model, the ESCO takes responsibility for both the performance and the credit risk. Therefore, the ESCO repays the loan and undertaking the credit risk; then the client does not take the financial risk. This model is shown in Figure 6.4. In developing countries, ESCO markets like Thailand and China, the shared savings model is more suitable since it does not require clients to take the responsibility of investment repayment risk.

ESMAP Finance Structure

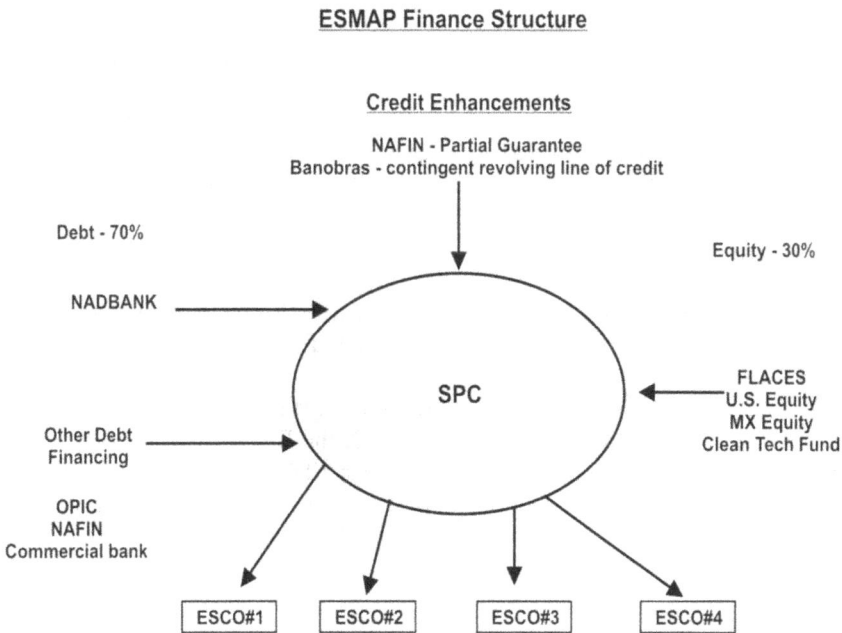

FIGURE 6.3 Energy Sector Management Assistant Program (ESMAP) with special purpose company (SPC).

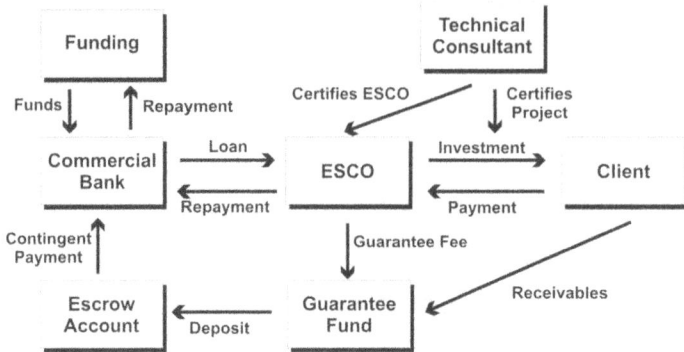

FIGURE 6.4 Basic model of ESCO with the involvement of a guarantee fund.

6.6.2 GUARANTEED SAVINGS MODEL

In this type of a model, the financial agent (bank) finances the client directly, as it is better capable of assuming any financial risks than the ESCO. The loan is to be paid back by the client to the bank in return for the financial help provided (Dreessen, 2003).

There are two points of discussion, as follows:

1. If energy savings are less than guaranteed energy saving by the ESCO the difference of the loan amount and the energy savings will be paid to the bank.
2. If energy savings are more than guaranteed energy saving by the ESCO the difference of the approximate energy savings and the actual saving made that excess amount will be paid by client (Dreessen, 2003).

6.6.3 SHARED SAVINGS MODEL

The shared savings model means that ultimately, both the credit risk and performance risk will be taken by the ESCO to assure that the suggested plan is working smoothly and it will be responsible for loan payment. If the revenues of the client stop, then it is the ESCO which must repay the loan. If the actual energy savings are more than what was approximate by the ESCO, the ESCO minimizes the energy savings and takes away more of the excess savings. ESCOs provide a detailed review of the energy consumption, including property of the customer and providing a plan for increasing the energy efficiency (Vine, 2005; Bertoldi, 2014 and Dreessen, 2003).

6.7 RISK ANALYSIS

The implementation of ESCO will give rise to risk factors which may be due to any one or both of the technical and financial risk. The risks which are mitigated or addressed will affect the desired financial returns. Risks can be grouped into three key issues: commercial, financial and technical risk. Technical risk is related with performance of ESCO to distribute the benefits of its proposed technical design and application. Table 6.3 provides a general perspective of the minimum risk considerations to be taken into account when evaluating an ESCO under these three categories.

6.7.1 TECHNICAL RISK

6.7.1.1 Engineering Design

ESCO benefits are very much dependent on operating conditions of design, and design criteria like production, capacity, load, operation, maintenance and all parameters are sometimes difficult to analyze or verify once the ESCO has been installed. Design reliability, equipment technology and the service provider are for consideration by the ESCO.

TABLE 6.3
Holistic Technique of the Minimum Risk Considerations

1. Technical Risk	Is the engineering design suitable or not?
1a. Engineering	What type of modifications/changes are required?
design	Do operating conditions meet criteria onsite?
	Are measures to be taken for reliability of the design? What is the probability of design failure?
	What are the pros and cons of ESCO failure on the business activity?
1b. Implementation	What are the physical constraints in the implementation or construction work?
	Is there any construction delay in the project?
	Are safety and health measures in place?
	Are the design specification and quality of ESCO up to the benchmarks, or not?
1c. Operation	Will the ESCO operate according to the design specification?
maintenance	Are there appropriate resources to operate and maintain the ESCO?
	How are benefits to be monitored and verified?
	What type of support is required in case of failure?
	What are the affects of failure to the main business?
	What is the expected time for corrective actions?
2. Commercial risk	Will the cost be subject to change?
2a. Procurement	Are all parties meeting their respective contractual obligations?
	Will the ESCO be locally procured?
	Is the cost open to exchange fluctuations?
2b. Counterpart	Is the manufacturer, supplier or service provider financially stable?
	How many counterparties are there to deal with on procurement?
	What is the risk of the counterparty defaulting on its obligations?
3. Financial Risk	
3a. Expected Returns	Are the obstacles rates appropriate to the ESCO?
3b. Financial structure	Does the ESCO present the best returns on financial resources?
	Is the ESCO related to the current and long-term financial plan?

6.7.1.2 Implementation

There are many factors due to complexity and scale by which ESCO implementation will be affected. There are always plenty of difficulties in building and construction works are delayed due to unprecedented situations onsite (site difficulties, weather, quality and safety), which may affect overall business performance. Work permits are required from authorities, which will cause significant delays in the implementation

Implementation of ESCOs require technical schedules, as they govern the quality delivered and expected performance of the ESCO.

6.7.1.3 Operation and Maintenance (O&M)

Design is critical, and operation and maintenance (O&M) predominates if the benefits of the design of the ESCO can be accomplished overall its operational life cycle. If benefits are undermined, the expected benefits may decrease, or in some cases operational costs increase. Dedicated resources with the technical expertise is required.

6.7.2 COMMERCIAL RISK

6.7.2.1 Procurement

In procurement processes, the most frequent risk which is mitigated is the variation of the cost after the investment decision. ESCOs manage through multifold parties and contracts, adding risk to the non-delivery in reference of contractual obligation. This process requires legal support, which increases the cost burden of the ESCO. If the ESCO is sourced outside the country cost will be increased due to foreign exchange fluctuations, which increases the price for ESCOs with prolonged delivery times and payment intervals.

6.7.2.2 Counterparty Risk

Operation and maintenance, and their implementation of the ESCO, depends on the service provider or supplier/distributor, according to its obligation of contracts. Financial stability is key issue in the ESCOs with complexity and a long life cycle.

6.7.3 FINANCIAL RISK

6.7.3.1 Expected Return

ESCO investment is established with the risk associated with the investment class; holistically, the higher the risk with the ESCO, the higher the expected rate of return. If all the associated risks are taken into consideration, investments in ESCOs are perceived to carry a much higher risk than investment types or classes.

6.7.3.2 Financing Structure

Financial structure is one of the important issues and hurdles to mitigation of ESCOs. Additional loans, security loans or debt on an balance sheet of enterprise increases the risk of the industries which may be in opposite to the shareholders' interests.

6.7.4 RISK PERCEPTION AND MITIGATION

The mitigation of risk is highly biased because it is based on inspection. The risk evaluation is based on perception, and if perception differs, than it is open to different opinions and interpretations. Liquidity is the paramount issue which prevents ESCO uptake in all domains. Risk is mitigated through actions of prevention. The outstanding risks—like the failure of equipment, and accidents—cannot be undermined, and it should be covered by some type of insurance. Henceforth, implementation of ESCOs may provide environmental benefits in a positive direction. On the contrary, investment in ESCOs should fundamentally reduce the risk profile. However, highly overconfident and aware financial providers do not recognize this when it comes to financing ESCOs.

6.8 BARRIERS TO IMPLEMENTING ESCOs IN PROMOTING ENERGY EFFICIENCY

Energy efficiency is multifaceted, focusing on issues concerning legal frameworks, economic and financial incentives, information, technology and knowledge gaps. There are numerous barriers faced in both develop and developing countries in executing ESCOs for energy efficiency. As described in what follows, these barriers can be grouped into (i) financial and (ii) economic knowledge and information.

- Deficiencies of policy, regulatory and legal enforcement.
- Deficiencies in support of government.

6.8.1 INSTITUTIONAL BARRIERS

- Institutional capacity is limited in public and private.
- Diversity among end-users for breakup of energy consumption and business models.
- Standardization is not up to the standards.

6.8.2 FINANCIAL AND ECONOMIC BARRIERS

- Pricing of energy and cost.
- Market trends and structure.
- Scarcity of financial incentives.

6.8.3 KNOWLEDGE AND INFORMATION BARRIERS

- Lack of priorities and awareness of energy efficiency programs.
- Lack of knowledge and awareness about ESCO operation.
- Lack of education and training.
- Scarcity of infrastructure.
- Lack of availability of energy efficiency technology.

6.9 CONCLUSION

ESCO is executed on a wide scale in many countries across the globe. The development of ESCOs is very predominant and there are significant factors that will affect ESCOs, such as energy prices, environmental issues, financial policies, international policies and markets. Even with the same ESCO policies implemented in several countries, the results could be different. The implementations of ESCO are required to adjust with the market participation and need. Several key points based on above discussions are placed below.

- Japan, Germany and the UK have the large markets, and similarly developed countries such as Austria have efficient market policies for small and medium enterprises (E. C.-J. R. Center, *European ESCO Market Survey*, 2012; LLC, 2013).
- Institutional organizations should create their own ESCO associations to overcome weakness. The ESCO association assists with implementation of promotional policies to disseminate information and assist governments with policy development.
- The shared savings model is used and implemented in every country.
- Implementation of pilot projects through government policies and identifying markets is important when designing the correct ESCO policy.
- Local bank financial activities and promotion of ESCOs to improve the financial mechanism is the need of the hour for energy efficiency in developing countries.
- International organizations have crucial roles and responsibilities for financial promotion of ESCOs, because the funding mechanism is difficult and complicated.
- Regulations and standards should be increased to improve the EPC, and ESCO accreditation is important.
- Loan guarantee policies for each country will be the financial solution, since this will provide security to the financial institutions to finance the projects.

REFERENCES

Bertoldi P., Boza-Kiss B., Panev S., and Labanca N. (2014); *ESCO market report 2013*, European Commission, Luxembourg.

Bertoldi P., Rezessy S., and Vine E. (2006); "Energy service companies in European countries: Current status and a strategy to foster their development," *Energy Policy*, vol. 34, pp. 1818–1832.

C.C.A.P. (CCAP). (2012, 9 October 2016); *Revolving and esco funds for renewable energy and energy efficiency finance: Thailand*. Available: http://ccap.org/assets/CCAP-Booklet_Thailand.pdf

Center E. C.-J. R. (2012); *European ESCO market survey*. Available: https://www.surveymon-key.com/s/JRC-European-ESCO-Survey

Da-li G. (2009); "Energy service companies to improve energy efficiency in China: Barriers and removal measures," *Procedia Earth and Planetary Science*, pp. 1695–1704.

Dreessen T. (2003); *Advantages and disadvantages of the two dominant world ESCO models; shared savings and guaranteed savings*, proceedings First Pan-European Conference on Energy Service Companies, Milan.

Ellis J. (2009); *Energy service companies in developing countries: Potential and practice*, International Institute for Sustainable Development (IISD), Winnipeg.

The Energy Service Market (2020); M. I. LLC 2013.

European Commission (2016, 3 October 2016); *Energy performance contracting.* Available: http://iet.jrc.ec.europa.eu/energyefficiency/european-energy-service-companies/energy-performance-contracting

Hansen S. J. (2003); *Lesson learned around the world*, presented at the First European Conference on Energy Service Companies (ESCOs): Creating the Market for the ESCOs Industry in Europe.

Hansen S. J. and Associates. (2011); *ESCOs around the world*, vol. 30, River Publishers, New York.

IEA (2015); *World energy outlook*, International Energy Agency, Paris.

Lee M.-K., Park H., Noh J., and Painuly J. P. (2003); "Promoting energy efficiency financing and ESCOs in developing countries: Experiences from Korean ESCO business," *Journal of Cleaner Production*, vol. 11, pp. 651–657.

Marino A., Bertoldi P., and Rezessy S. (2010); *Energy service companies market in Europe*, European Commission, Luxembourg.

Morgado D. (2014, October 4); *Energy service companies and financing.* Available: https://www.iea.org/media/training/presentations/latinamerica2014/8A_Energy_Service_Companies_and_Financing.pdf

Murakoshi C. and Nakagami H. (2009); *Current state of ESCO activities in Asia: ESCO industry development programs and future tasks in Asian countries*, European Council for an Energy Efficient Economy, Stockholm.

Okay N. and Akman U. (2010); "Analysis of ESCO activities using country indicators," *Renewable and Sustainable Energy Reviews*, vol. 14, pp. 2760–2771.

Seefeldt F., Offermann R., Duscha M., Brischke L.-A., Schmitt C., Irrek W., et al (2013); *Marktanalyse Und Marktbewertung Sowie Erstellung Eines Konzeptes Zur Marktbeobachtung Für Ausgewählte Dienstleistungen Im Bereich Energieeffizienz*, Prognos, Berlin.

Taylor R. P., Govindarajalu C., Levin J., Meyer A. S., and Ward W. A. (2008); *Financing energy efficiency: Lessons From Brazil, China, India, and beyond: The international bank for reconstruction and development*, The World Bank, Washington, DC.

Vine E. (2005); "An international survey of the energy service company (ESCO) industry," *Energy Policy*, vol. 33, pp. 691–704.

World Bank (1 October 2016); *Fostering the development of eSCo markets for energy efficiency.* Available: http://documents.worldbank.org/curated/en/709221467753465653/pdf/103932-BRI-LW54-OKR-PUBLIC.pdf

Websites:

[1]   https://beeindia.gov.in/latest-news/latest-grade-wise-list-empanelled-escos-bee

7 Energy Management and Monitoring Systems

Jitendra Saxena

7.1 INTRODUCTION

The best example to discuss about energy management is our human body. The body as a whole can be treated as an energy system in which different organs can be similar to the parts of industrial energy process systems. Each part of the energy system is important to maintain our intrinsic and extrinsic energy. In order to regulate our body, each sub-system should be efficient. The whole human body requires regular monitoring with nutrition and medication. Similarly, process systems require periodic monitoring and follow up for efficient use of energy [1 and 2].

The effective utilization of energy use for maximizing profit and minimising energy costs can be attained by optimum utilization of energy sources for growth and sustainability. The energy monitoring, statistics, energy use and its consumption plays a vital role for energy management in metal industries.

7.2 OBJECTIVES OF ENERGY MANAGEMENT AND MONITORING SYSTEMS

From the definition, purpose and procedures of energy management and audit as explained in Chapters 3, 4 and 5, it is obvious that the objectives of the same can be fulfilled on a sustainable basis only through an appropriate monitoring system for the metal industries.

7.3 BENCHMARKING AND ENERGY PERFORMANCE

Benchmarking is a very vital and powerful technique for performance analysis and assessment. Energy consumption and cost trends with historical data are provided month-wise and day-wise. Energy consumption and SEC with trend analysis supports analysis of the impact of capacity utilization and use of energy efficiency and costs on macro level to assist in optimization. (Energy Audit Reports of National Productivity Council 2017-2018)

7.3.1 Plant Energy Performance

Plant energy performance is a comparative study of one year with reference to other years, considering production output. Performance monitoring of energy provides overall study of energy usage of a plant at a reference year with the subsequent years to evaluate and improve plant performance.

A plant's production output changes every year and the output is dependent on availability of resources including energy, plant's availability factor, market demand and policy of the organization. The desired energy is evaluated to create production output of this year if the plant had the same operating conditions as compared with the reference year. The actual value is compared to determine the significant improvement that has in use since the reference year.

7.3.2 PRODUCTION FACTOR

Production is the ratio of production in the current year to that in the reference year.

Production factor = (Current year's production) / (Reference year's production)

7.3.3 REFERENCE YEAR EQUIVALENT ENERGY USE

The current year's production is obtained in relation to reference year energy use, and the output may be called the "reference year energy use equivalent" or "reference year equivalent" in technical terms. The reference year equivalent is obtained by multiplying the reference year energy use by the production factor.

Reference year equivalent energy use = (Reference year energy use) × (Production factor)

7.3.4 OPTIMIZING THE INPUT ENERGY REQUIREMENTS

The range of measures for optimizing the input energy requirements are:

- Matching of compression operational needs in parallel or series.
- Insulation thickness periodic checkup and review.
- Transformer loading optimization techniques and implementation.

7.3.5 FUEL AND ENERGY SUBSTITUTION

Fossil fuel substitution relates to requirements to match with less costly and more efficient fuels such as biogas, natural gas and agro-residues. Energy usage and its effective utilization are the vital parameters for production. Dependency on energy can be reduced by energy conservation and substitution. The need and requirement of natural gas is increasing at exponential rate as fuel and feedstock in the petrochemicals industries, power sector, and sponge iron industries. A few examples of fuel substitution are listed in what follows.

1. Coal replacement by rice husks, coconut shells, etc.
2. Liquid diesel oil replacement by low sulphur heavy stock (LSHS).
3. Electrical heaters replacement by steam heaters.
4. Steam-based hot water systems replacement by solar-based systems.

7.4   KEY ELEMENTS OF MONITORING AND TARGETING SYSTEMS

Monitoring and targeting (M&T) are techniques of energy management in which all plant industrial utilities are managed efficiently with cost minimization. Input energy sources like steam, refrigeration, fuel, effluent, water, compressed air and electricity are managed and controlled efficiently in the same pattern that raw materials, finished product inventory, buildings and capital are managed.

The important elements of an M&T system are listed in what follows.

- Recording: annual energy consumption, monthly energy consumption or daily energy consumption should be recorded and monitored systematically.
- Analysis: correlation of annual energy consumption to a measured desired output, such as production quantity and load factor, is determined.
- Comparing: energy consumption is compared to a desired benchmark or standard.
- Setting Targets: energy consumption is to be controlled by setting targets.
- Monitoring: energy consumption comparison is observed and targets set.
- Reporting: results of variances to be reported from the targets which have been set.
- Controlling: correction in variances which may have been identified and needs to be controlled.

Particularly, an M&T system will involve the following.

- Checking: energy invoices are to be checked and physically verified.
- Allocating: energy accounting centers are to be identified and allocated for energy costs.
- Determining: energy performance or efficiency should be determined holistically.
- Recording: energy efficiency and energy use is to be checked.
- Fixing: problems in systems and equipment are to be highlighted and rectified.

7.4.1   PRINCIPLES FOR MONITORING, TARGETING AND REPORTING

The key performance indicators for the energy used for any business will vary with production process and input. Key performance indicators of energy use will provide the following.

- Determination if current energy consumption is greater than or less than before.
- Variations of energy use by changing controllable aspects of your business.
- Energy waste identification for its control and optimization.
- Business comparison and benchmarking processes.
- Reaction of business changes in the past.

- The energy management program and its performance targets/external benchmarks are:
  - Operational range.
  - Vintage of technology.
  - Quality and specifications of raw materials.
  - Quality and specifications of products.

Benchmarking energy performance permits the following.

- Quantification of fixed and variable energy consumption patterns and different production levels.
- Industry energy performance comparative studies with respect to various production levels (capacity utilization).
- Best practices identification (on the basis of external benchmarking data).
- Scope and margins available for energy consumption and cost reduction.
- A basis for monitoring and target setting exercises.

The benchmark parameters can be the following.

- Gross production related to hot metal produced expressed as kWh/MT parameters used in iron and steel plants; heat rate of a captive power plant in the metal industries expressed as kCal/kWh and liquid metal output in a foundry as kWh/MT.
- For a power plant/cogeneration plant—plant percentage of loading, condenser vacuum and inlet cooling water temperature would be important factors to be mentioned, along with heat rate (kCal/kWh).

For a foundry unit, furnace type and melt output are very important parameters, composition (mild steel, high carbon steel, cast iron, etc.), raw material mix and peak current generated during arcing of electrodes in electric arc furnace are the most useful and significant operating parameters to be reported while monitoring SEC data (Table 7.1).

TABLE 7.1

Monitoring of Technical Parameters in Rotary Hearth Furnace in a Secondary Steel Plant in which Slabs Are Reheated for Downstream Rolling and for Heat Balance

Parameters of Study/Monitor/Calculation	Units/Remark
Calculation of Heat Input	
Calculation of Heat Output	
Heat Balance Table	
Duration of trial	Hours (i.e. 16 hrs, 20 hrs or 24 hrs)

Parameters of Study/Monitor/Calculation	Units/Remark
Hydrogen in fuel	Percentage
Water content	Percentage
Calorific value of LSHS	10,200 kCal/kg
Specific gravity of LSHS	0.95
Specific heat of fuel	0.5kCal/kg°C
Fuel consumption	In liters
Theoretical air required for combustion	kg of air/kg of fuel
Temperature of fuel before firing	°C
Temperature of furnace	°C
Input material	Tons
Production	Tons
Scale loss	Percentage
Specific heat of steel at ambient temperature	kCal/kg°C
Specific heat of steel at furnace temperature °C	kCal/kg°C
Heat of formation of scale	kCal/kg of iron
Humidity of air	kg/kg of dry air
Atomizing air	Nm³/hr
Secondary air temperature	°C
Specific heat of air at secondary air temperature	kCal/kg°C
Oxygen in flue gas	Percentage
Flue gas temperature after air pre-heater	°C
Specific heat of flue gas	0.26 kCal/kg°C
Specific heat of water vapor	0.48 kCal/kg°C
Cooling water flow rate	m³/hr
Temperature difference	°C
Number of doors	2 nos
Factor of radiation	0.7 for rectangular door
Area of opening (A)	m²
Diameter of furnace	M
Height of furnace	M
Wall surface area	3.14DL
Roof Surface area	3.14R²
Average wall temperature	°C
Average Roof temperature	°C
Calculation of production rate	T/hr
Calculation of fuel rate	Liters/hr
Calculation of specific fuel consumption	kg/T of billet
Calculation of specific atomizing air	kg of air/T of billet
Calculation of specific secondary air	kg/T of billet
Calculation of specific flue gas	(Air + fuel) in kg/T of billet
Calculation of specific water vapor	$(9H_2\% + M\%) / 100$
Calculation of specific scale formation	kg/T of billet
Calculation of specific cooling water	kg/T of billet

1. Combustion heat of fuel
2. Sensible heat of fuel
3. Sensible heat of atomizing air
4. Quantity of heat brought in by charged steel
5. Heat of formation of scale

(*Continued*)

TABLE 7.1 Continued

Parameters of Study/Monitor/Calculation	Units/Remark
Total heat input	a + b + c + d + e kCal/T of billet
1. Sensible heat of steel	Total heat output = a + b + c + d + e + f + g
2. Sensible heat of scale	
3. Sensible heat of dry flue gas	
4. Heat loss due to formation of water vapor from fuel	
5. Heat loss due to formation of water vapor from air	
6. Heat taken away by cooling water	
7. Calculation of radiation heat loss	
Heat input in kCal/T	Heat output
Combustion heat of fuel	Sensible heat of steel
Sensible heat of fuel	Sensible heat of scale
Sensible heat of atomizing air	Sensible heat of dry flue gas
Quantity of heat brought in by charged steel	Heat loss due to formation of water vapor from fuel
Heat of formation of scale	Heat loss due to formation of water vapor from air
Total = XYZ kCal/T and 100%	Heat taken away by cooling water
	Radiation heat losses
	Others
	Total = XYZ kCal/T, 100%

Case Study 7.1: Steel Plant Specific Energy Consumption for Base Year and Current Year

In a steel plant, daily sponge iron production is 600 tons. The sponge iron is further processed in a steel melting shop for production of ingots. The yield from converting sponge iron into ingots is 90%. The plant has a coal-fired captive power station to meet the entire power demand of the steel plant. The base year (2019) and current year (2020) energy consumption data are given in Table 7.2.

TABLE 7.2
Energy Consumption Data for Base Year 2019 and Current Year 2020

Parameters	Base Year (2019)	Current Year (2020)
Sponge iron production	600 T/day	600 T/day
Specific coal consumption for sponge iron production	1.5 T/T of sponge iron	1.4 T/T of sponge iron
Specific power consumption for sponge iron production	150 kWh/T of sponge iron	140 kWh/T of sponge iron
Yield in converting sponge iron into ingot in steel melting shop	90%	90%

Parameters	Base Year (2019)	Current Year (2020)
Specific power consumption in steel melting shop to produce ingots	1,000 kWh/T of Ingot	900 kWh/T of Ingot
Captive power station heat rate	4,000 kCal/kWh	3,800 kCal/kWh
GCV of coal	5,500 kCal/kg	5,200 kCal/kg

The solution is shown in Table 7.3 and Table 7.4.

TABLE 7.3
Base Year (2019) Plant-Specific Energy Consumption

Specific energy consumption for sponge iron	1,500 kg × 5500 + 150 kWh × 4,000 = 8.88 MkCal/T
Total energy consumption for base year	8.88 × 600 = 5,328 million kCal/day
Actual production, considering 90% yield from sponge iron to ingot conversion	600 × 0.90 = 540 tons/day
Specific energy consumption for ingot	1,000 × 4,000 = 4 million kCal/ton of ingot
Total energy consumption for ingot production per day	4 × 540 = 2,160 million kCal/day
Plant-specific energy consumption for production of finished product (ingot) during base year	(5,328 + 2,160) / 540 = 13.86 million kCal/ton

TABLE 7.4
Current Year Specific Energy Consumption (2020)

Specific energy consumption for sponge iron	1,400 × 5,200 + 140 × 3,800 = 7.81 million kCal/ton
Total energy consumption for base year	600 × 7.81 = 4,686 million kCal/day
Actual production, considering 90% yield from sponge iron to ingot conversion	600 × 0.90 = 540 tons/day
Specific energy consumption for ingot	900 × 3800 = 3.42 million kCal
Total energy consumption for ingot production per day	540 × 3.42 = 1846.8 million kCal/day
Plant-specific energy consumption for production of finished product (ingot) during base year	(4686 + 1846.8) / 540 = 12.09 million kCal/ton

7.5 ENERGY ANALYTICS WITH APPLICATION OF AI AND IOT

Artificial Intelligence (AI) gives knowledge of different technologies to track functions of human brains such as reasoning, perception and learning, and making effective decisions through algorithmic processing and programming. Machine learning is the effective applications of protocols, instructions and algorithm development to analyze and interpret data and learn from data, then making effective decisions and predictive analysis of real-time global events. Deep learning is a rare and specific branch of machine learning. It is the process by which data is to be processed for simulating and building the human brain for analysis.

The bottlenecks of traditional machine learning break through the deep learning approach and what is not achieved by traditional machine learning algorithms. Computer vision and its application for algorithm development to detect and identify targets in videos and images, to measure and detect them, includes the following levels.

7.5.1 PLANT LEVEL

Plant layout software and hardware design schematics with process and procedures used to collect metered data that can be collected in a variety of methods ranging from the following.

1. Taking plant meter readings by walkthrough audit and then making a database of readings in a Microsoft Excel spreadsheet.
2. Data capturing and data histories can range from spreadsheets to proprietary databases which are structured query language (SQL) and non-structured query language (non-SQL) as database enrichment takes place through monitoring energy guzzlers, detailed audit, post-audit performance monitoring, etc.

7.5.2 PLANT DEPARTMENTAL LEVEL

The comparison of energy consumption and its process with the plant utility department allows arrival at results on energy performance analysis. Data analysis may be described as the technique of converting data into management information. Energy consumption data—daily energy consumption, monthly energy consumption and annual energy consumption—assist in establishing a pictorial view of the industry and helps provide realization on how to focus on energy conservation strategies.

7.5.3 SYSTEM OBJECTIVE LEVEL

To develop a system objective level, analyze and assess a proposed energy management system (EMS) objective level and focus on an energy management information system (EMIS) perspective and business studies. This consists of the following.

1. Industrial organizational facility gaps and views of industrial organizations as per the recommendation of the EMIS.
2. EMIS with vision and mission.
3. The business case studies for investing in EMIS, including predictable costs and energy savings.

7.5.4 EQUIPMENT LEVEL

An outline for the business model of the proposed EMIS system is to be developed with a work profile based on recommendations and assessment of the EMIS audit report. At this point, that sign-off takes place from management and key decision-makers on the design of the EMIS structure. Detailed work packages are created with conceptual designs built up for all EMISs (i.e., metering, data analysis, data capture, etc.). Bids can be acquired by external service providers for this type of specific work.

7.6 ENERGY MANAGEMENT INFORMATION SYSTEMS AND ISO 50001

An EMIS gives overall energy performance of different plant sectors, enabling departments to take proper action to manage energy efficiently. The results are improvements in productivity through the completion of energy performance and energy savings opportunities for long-term strategies. The business case performance metrics will inform the decision to proceed or not with the EMIS implementation plan phase. An implementation plan is a structured guide to provide guidance, tools and activities at each phase of EMIS implementation planning. The final part includes a reference document which reproduces the EMIS, achieving improved energy efficiency.

ISO 50001 was adopted in 2011 (latest version released in 2018) and developed by experts of international repute, and its application is adopted universally. It has been globally adopted more by energy intensive metal industries in the present times since its launch in comparison to two other popular standards: Quality Management ISO 9001 and Environmental Management ISO 14001. ISO 50001 (2011) provides the following framework of requirements for organizations.

- Policy for effective and judicious use of energy.
- Setting of targets to achieve the policy.
- Energy usage of data and making corrective decisions about the use of energy.
- The outcomes/results are to be measured.
- Improvement in energy management by 360°.

ISO 50001 starts with a high-level strategic management and will provide activities of operational nature such as monitoring and targeting of energy use. It provides an international and systematic approach to best practices which can be applied universally and in all industrial sectors for continuous improvement in energy performance of metal industries through application of P-D-C-A (Plan-Do-Check-Act) cycle.

7.7 APPLICATION OF AI IN METAL INDUSTRIES AND ITS IMPROVEMENT

Exponential growth has recently been achieved in the mathematical groundwork of AI: soft computing. The fundamental concepts of soft computing are now being included in the methodology of industrial automation. Significant growth in the applications of AI systems in the industrial processes—and systems including metal and metallurgy—are being forecasted for the near future (Chertov, 2003; Ritamaki and Luhtaniemi, 1999).

Metallurgical production provides a huge collection of composite distributed operations, from preparation of raw materials to manufacture of metal products. Each operation is characterized by a certain degree of imprecision (uncertainty). It is in such cases that the use of soft computing is effective. It can be said with confidence that soft computing has already been responsible for a great deal of development in the metallurgical industry ([16]; Ujjwal Bhaskar et al., 2003).

The melting process of metal with its control—as explained in Figure 7.1 with an aim to obtain validated results from previous practical studies and industrial performance, which are related to white cast iron and its chemical composition—are being analyzed by application of the machine learning algorithm. The machine learning algorithm gives a complete idea covering subjective knowledge and performs prediction analysis for the chemical composition of white cast iron. The foundry-collected data are entered into the inputs and outputs for the development of the machine learning algorithm. The input dataset for the machine learning algorithm to be trained consists of a chemical composition of 2.5 tons of molten metal in the furnace, five tons of steel waste chemical composition and a final composition of molten metal (Rendueles et al., 2001).

The training machine learning algorithm output dataset is the amount of additives alloy that are added during the alloying process (Ujjwal Bhaskar et al., 2003; Peng et al., 2000).

Figure 7.2 provides a detailed design schematic of a neural network (NN-12–16–4) has 16 neurons with one hidden layer in it, the input layer is defined by the number of outputs which are four (4). The neurons in input and hidden layers of neural networks had the sigmoid transfer function while the neurons of the output layer have the linear transfer function.

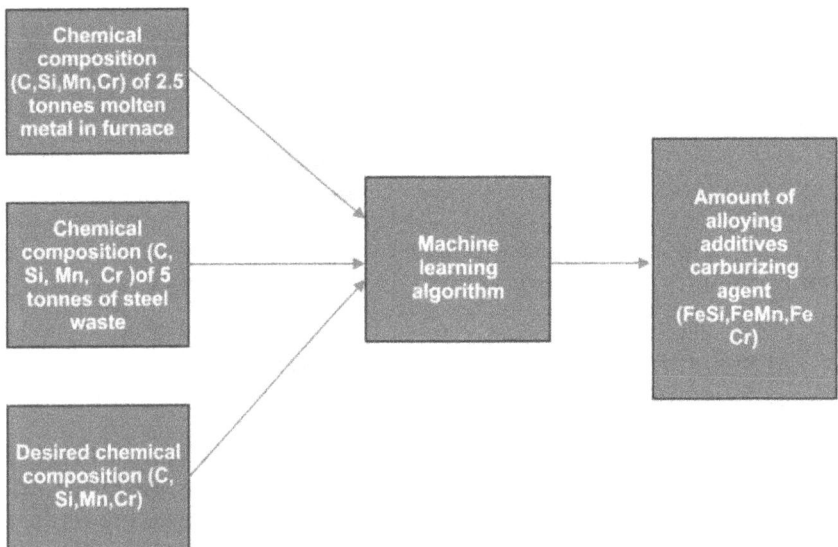

FIGURE 7.1   Control of white cast iron production.

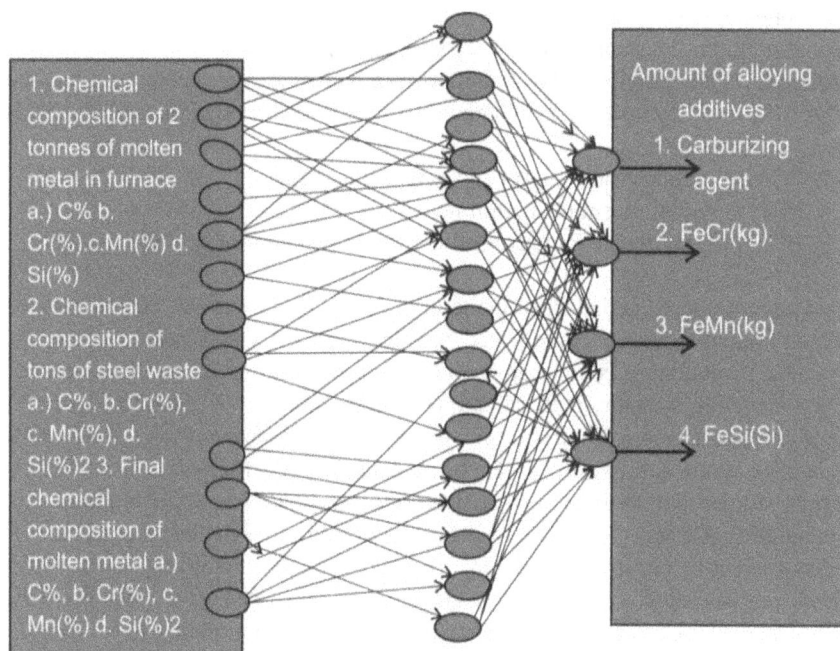

FIGURE 7 2 Neural network architecture for prediction of the amount of alloying additives.

TABLE 7.5

Maximum and Mean Error in the Testing Phase of the Neutral Network

Error Type	Carburizing Agent (kg)	FeCr (kg)	ГeMn (kg)	FeSi (kg)
Mean error (%)	0.47	0.51	3.31	1.88
Maximum error	1.97	1.92	9.42	8.18

Table 7.5 provides mean square error results in the training phase as a very significant performance of neural network as shown in Figure 7.2 (NN12–16–4) is 0.16. NN 12-16 -4 means neural network with one hidden layer with 16 neurons, 12 inputs and 4 outputsAs a validity measure of a neural network, maximum error and the mean errors in the test phase are used. The testing has been performed with a set of 20 data points that did not participate in the development of the neural network model. The maximum and mean errors for the prediction of four alloying additives are given in the table.

The roadmap taken by AI from the professor's classroom to the factory floor has been a lengthy one. Very large–scale integration (VLSI) is a specific field of engineering that encompasses improving the sensitivity of transducers, emergent open technologies that integrate different computer platforms, and related issues.[1] VLSI devices make it possible to realize the vast computing potential of microprocessors in real

1 Appl. Sci. **2020**, 10, 6048important doi:10.3390/app10176048 www.mdpi.com/journal/applsci

time. For example, VLSI has made real-time control in the interactive management a feasible solution. It has also been used to develop exclusive transducers that not only introduce AI into peripherals, but also significantly increase the speed of control operations which employ task delegation and distributed intelligence (Tautz et al., 1997).

1. SCADA: supervisory control and data acquisition.
2. ERP: enterprise resource planning.
3. PLC: programmable logic controller/ distributed control system (DCS).

Rapid advancement that has been made in the areas of fuzzy logic, neural networks and soft computing has become a leading trend in industry and includes the following fields: evolutionary computing, genetic algorithms, artificial intelligence, chaos theory modeling systems, probabilistic reasoning, study of AI control algorithms, image and pattern recognition, study of the principles of mechanisms self-adaptation of complex systems, fuzzy databases, fuzzy data searches, fuzzy sets, etc. Literature surveys, studies and findings in these emerging areas have found major applications in metallurgy. Definite successes have been achieved in particular in predicting the silicon content of pig iron during blast furnace smelting, monitoring laser welding operations, checking the quality of welds, finding hidden defects in rails, predicting the hydrogen content of steel during vacuum degassing, optimizing hot rolling and monitoring the deformation of the semi-finished product during rolling, modeling the solidification of a continuous-cast semi-finished product, monitoring the change in the temperature of the strip during hot rolling and monitoring the thickness of the coating on steel in a hot galvanizing line, modeling cold rolling checking the level of the steel in a tundish during continuous casting, etc. [3].

It has been nearly a half century since the first AI models were constructed. AI was originally based on strict, precise computing. Later, when real-world inaccuracies were taken into account, researchers' emphasis shifted to soft computing. The reason for the paradigm shift was an understanding that traditional exact binomial logic systems and studies in set theory and the theory of probability are inadequate to the task of dealing with imprecision, uncertainty, and the complexity of the real world. Soft computing is based on neural networks (NNs), fuzzy logic (FL) and probabilistic reasoning (PR) (Wang and Tieu, 1997).

The functions of NNs are to examine the degree of correspondence between sets of data and to identify parameters, images, and systems. FL deals with imprecision and probabilistic descriptions of processes. Recently, FL has been merging with information genetics. It is now being used to evaluate uncertainty, systematize random searches and aid in optimization.

7.7.1 INDUSTRIAL APPLICATIONS IN THE METAL INDUSTRIES OF FUZZY LOGIC AND NEURAL NETWORKS

7.7.1.1 Galvanizing and Coating Lines

The thickness of coating applied to finished products on galvanizing lines is controlled and monitored by the weight of the coating based on the balance between the total weight of the metal and the weight of the electrolytes in the bath. Since the efficiency of the process is a nonlinear function of the input variables, the second are especially important for smooth process operation of the line. The efficiency of the galvanizing operation as a whole depends on the condition of the electrolytes. The optimum control point must be calculated in such a way that the values are constantly adapted to the actual process and take into account changes in the specified weight of the coating. Special monitors are installed in the line due to the substantial distance between the measurement gauge and the control transducers. The system reduces the consumption of metal for galvanizing to a minimum. The efficiency of the galvanizing line is based on the use of a fuzzy interactive system, and the interactive process by which the system is informed constantly adapts the mathematical model to the actual operating conditions of the equipment (Tautz et al., 1997; Rendueles et al., 2001).

7.7.1.2   Coke-Oven Batteries

The control system that is used to heat coke-oven batteries is continually being studied and improved. The first-generation system was a statistical model of heating (1987–1991). The operation of the second-generation (dynamic) system was based on the principle of calculating the energy balance. It employed feedforward and feedback control in determining the temperature at the end of the coking operation and was used during the period 1991–1995. The third-generation system was augmented by FL controllers. Their use made it possible to reduce the scatter of the end-point temperature to 15°C. The reliability of the system was also improved significantly (Tautz et al., 1997; Ritamaki and Luhtaniemi, 1999).

7.7.1.3   Cold Rolling

On a cold rolling mill, the thickness of the strip decreases in succession after each reduction. Thus, the control problem reduces to optimizing the reduction while minimizing the power associated with the load and satisfying certain restrictions. Since the properties of the materials and the friction coefficient are not precisely known, they are regarded as fuzzy numbers. Fuzzy set theory is used to determine the optimum reduction. A graph is usually used to attain the largest possible reduction in thickness after the initial passes, while minimizing the compressive load. However, the reliability of such controls is not great. Optimizing the process by using fuzzy set theory provides a solution which satisfies the requirements for reproducing reduction regimes with a high degree of reliability while minimizing the load (Ujjwal Bhaskar et al., 2003). Another fuzzy control system was developed to obtain the specified thickness of steel strip. The system consists of a fuzzy output controller and an adaptive controller with an NN output that performs a parallel modeling of the rolling operation. The fuzzy output controller generates the control signal, while the adaptive NN model prevents delays in generating the output signal and ensures that the structure and accuracy of the control operation are as prescribed. This control scheme works superbly (Dixit and Dixit, 2000).

7.7.1.4   Refining and Continuous Casting of Steel: Two Case Studies

In the oxygen converter shop of a factory in Great Britain (Port Talbot Basic Oxygen Steelmaking Plant), automated NNs predict carbon content of the molten steel and temperature at the end of the blow, and also predicting the consumptions of oxygen and coolants during the blowing operation. The data file used to inform the system about the course of the process consists of actual measurements of temperature and the carbon content of samples taken at the end of the blow. These measurements have been used as the input to an NN model in order to predict the consumption of additional oxygen or coolants and obtain the desired results at the end of the blow (Ujjwal Bhaskar et al., 2003; Ritamaki and Luhtaniemi, 1999; Dixit and Dixit, 2000; Wang and Tieu, 1997; Siemens). A factory in Italy (Danieli Automation) uses NNs to control continuous casting based on the level of the metal in the mold. In one variant, the system determines the level of the liquid metal and an FL controller corrects it (Cox et al., 2002; Watanabe et al., 1999). In another case, it was important to keep the level of the metal stable in order to ensure the necessary quality of the semi-finished product. It was difficult to use conventional methods to achieve the desired degree of control due to the nonlinear characteristics of the process. An NN model that is trained on the basis of a time series of the input parameters controls the operation of the stopper through the use of AI in the main control circuit (Michael et al.,; Liu and Zhao, 2000; Saarelainen et al., 2001; Ghaboussi et al., 1999; Peng et al., 2000; Wang and Tieu, 1997).

7.8   CONCLUSION

Energy management monitoring and analysis, with all the key issues of preliminary energy audit and finally detailed energy audit, will ultimately focus on monitoring and optimization of production processes and efficient operation with the latest technique of AI are explained in detail in this chapter. With the exponential rapid progress being made in industry worldwide on the introduction of fuzzy systems to control, optimize, monitor and analyze production processes, the metallurgical industry is developing a rational prospective on soft computing for improving product quality, process parameters and the efficiency of industrial operations. The proprietary research being done on soft computing was fundamentally customer-focused, and thus competitive by nature. However, it soon became clear that the best results can be achieved if there is some crossover from the different avenues of research. The effectiveness and validation of the analytical models with real-time data analysis of metal industries will be vetted by the use of fuzzy models which were explained in detail in this chapter.

REFERENCES

Bhaskar U., Kamal R., Kumar R. S., et al (2003); *Computerization and automation in the steel industry*, February 26–28. https://link.springer.com/article/10.1023/A:1027402528311
Chertov A. D. (2003); "Use of artificial in the intelligence systems metallurgical industry (survey)," *Metallurgist*, vol. 47, no. 7–8.
Chia G. W., Weber B., and Schlechter (2001); "The value of neural networks for galvanizing lines," *Siemens Mining and Metals Power*, pp. 246–251.
Cleaner Production (2016); *Energy efficiency manual prepared for GERIAP, UNEP, BANGKOK by national productivity council.* https://www.academia.edu/23727894/Cleaner_Production_Energy_Efficiency_Guidelines_for_the_Integration_of_Cleaner_Production_and_Energy_Efficiency

Cox I. J., Lewis R. W., Ransing R. S., Laxzczewski H., and Berni G. (1999/2002); "Application of neural computing in basic oxygen steel making," *Journal of Materials Processing Technology*, pp. 310–315, January 15.

Dixit P. M. and Dixit U. S. (2000); "Application of fuzzy set theory in the scheduling of a tandem cold-rolling mill," *Journal of Manufacturing Science & Engineering*, 494–500, August.

Energy Conservation (2017); *The Indian experience, department of power & NPC publication good practice manual in iron and steel sector of India (2017) energy audit reports of national productivity council*. Energy Conservation.

Ghaboussi J., Wu X., and Chou J.-H. (1999); *Detection of hidden faults in rails*, Summary of Engineering Research, University of Illinois, A Report on Activities During the Calendar Year, Urbana-Champaign.

Kusiak J., Leonard J. G., and Dudekl K. (2013); *Artificial intelligence approach to the modeling of rolling loads in technology design for cold rolling processes*, The Second International Conference on Intelligent Processing and Manufacturing of Materials, Honolulu, HI, U.S. IEEE, July, vol. 1. pp. 543–547.

Liu H. and Zhao Y. (2000); "Intelligent control of hot strip coiling temperature," *Journal of University of Science and Technology, Beijing (English Edition)*, pp. 147–150, June.

Michael J., Einar B., Bjorn F., and Thomas P. (199); "How neural networks are proving themselves in rolling mill process control," *Siemens Mining and Metals Power*, pp. 259–265.

Neural network optimizes scrap melting process (2001); "Neural network optimizes scrap melting process," *Siemens Mining and Metals Power*, pp. 98–99.

Peng L.-M., Mao X.-M., and Xu K.-D. (2000); "Simulation and control model for interactions among process parameters of directional solidification continuous casting," *Trans Nonferrous Metals Society of China*, pp. 449–452.

Rendueles J. L., Gonzalez J. A., Ortega F., and Vergara E. (2001); *Improvement in coating uniformity with an automatic neural setup system for air knives in a hot dip galvanizing line*, 5th International Conference on Zinc and Zinc Alloys, Galvatech.

Ritamaki O. and Luhtaniemi H. (1999); *Advanced coking process control at Rautaruukki Steel*, Technical Exchange Session of the Committe on Technology, Washington, DC, December.

Saarelainen E., Kamarainen J.-K., Kalvainen H., and Vainola R. (2001); "Neural prediction of hydrogen in vacuum tankdegassing," *Iron and Steelmaker*, pp. 55–58, March.

Tautz W., Wilke E., and Goepel J. (1997); *Coating weight fuzzy control for electrolytic galvanizing lines*, Flat Product Technology, Seoul, Korea, May 12–14.

Wang S. and Tieu A. K. (1997); *Enhanced fuzzy interface control for strip thickness in a rolling mill*, International Conference on Thermomechanical Processing of Steels and Other Materials, Wollongong, Australia, July 7–11.

Watanabe T., Omura K., Konishi M., et al (1999); "Mold level control in continuous caster by neural network model," *ISIJInt*, pp. 1053–1060.

Websites:

[1]   https://www.npcindia.gov.in/NPC/User/Competencies_EnergyManagement

[2]   "The optimization and methodology of energy design using energy systems and process so as to reduce energy requirements per unit of output." (http://www.npcindia.org.in/wp-content/uploads/2014/07/GHG-Manual-Iron-Steel-Final.pdf)

[3]   https://www.osti.gov/etdeweb/ on Energy Technology Data Exchange (ETDEWEB)

8 Sustainability in Metal Industries

Binoy Krishna Choudhury

8.1 INTRODUCTION

Sustainability is a challenge for any creation. Metals have played an all important role in the past and will continue to do so in the foreseeable future of human civilization. Metal industries, too, face several challenges including sustainable supply of raw materials, energy and markets. In the long run, each of these is bound to be uncertain. The material world can never be stagnant. There must be either development or decline. Warnings of decline include resource shortages, water and land stress, environmental degradation and climate change—all are burning issues of modern times. These issues are adversely affecting sustainability of metal industries, as well.

The "official" definition of sustainable development was developed for the first time in the Brundtland Report in 1987: Sustainable development is the idea that human societies must live and meet their needs without compromising the ability of future generations to meet their own needs (United Nations, 1987). The question of sustainability has been getting increasing importance due to growing pressure on infrastructure (especially urban infrastructure), environment, resource allocation and utilization, use of energy among others caused by high growth rate of population and unorganized human settlements. As a follow up of Millennium Development Goals signed in September 2000 and expired in 2015, in the same year, the United Nations (2015) set 17 goals and 169 targets with a plan of action for people, planet and prosperity to be achieved within 2030. The 17 Sustainable Development Goals (SDGs) (Box 8.1) are integrated, indivisible and balance the three dimensions of sustainable development: the economic, social and environmental.

BOX 8.1   SUSTAINABLE DEVELOPMENT GOALS

The following 17 Sustainable Development Goals (United Nations, 2015) are integrated, indivisible and balance the three dimensions of sustainable development: the economic, social and environmental.

- Goal 1: end poverty in all its forms everywhere.
- Goal 2: end hunger, achieve food security and improved nutrition and promote sustainable agriculture.

DOI: 10.1201/9781003157137-8

- Goal 3: ensure healthy lives and promote wellbeing for all at all ages.
- Goal 4: ensure inclusive and equitable quality education and promote lifelong learning opportunities for all.
- Goal 5: achieve gender equality and empower all women and girls.
- Goal 6: ensure availability and sustainable management of water and sanitation for all.
- Goal 7: ensure access to affordable, reliable, sustainable and modern energy for all.
- Goal 8: promote sustained, inclusive and sustainable economic growth, full and productive employment and decent work for all.
- Goal 9: build resilient infrastructure, promote inclusive and sustainable industrialization and foster innovation.
- Goal 10: reduce inequality within and among countries.
- Goal 11: make cities and human settlements inclusive, safe, resilient and sustainable.
- Goal 12: ensure sustainable consumption and production patterns.
- Goal 13: take urgent action to combat climate change and its impacts.
- Goal 14: conserve and sustainably use the oceans, seas and marine resources for sustainable development.
- Goal 15: protect, restore and promote sustainable use of terrestrial ecosystems, sustainably manage forests, combat desertification, and halt and reverse land degradation and halt biodiversity loss.
- Goal 16: promote peaceful and inclusive societies for sustainable development, provide access to justice for all and build effective, accountable and inclusive institutions at all levels.
- Goal 17: strengthen the means of implementation and revitalize the global partnership for sustainable development.

Each of the 17 SDGs are pertinent to metal industries; however, more emphasis may be given to achieve SDGs 6, 7 and 12–15. The huge scope of sustainability is, according to some schools of thought, justified in industrial scenarios with the help of three verticals: environmental sustainability, social sustainability and governance sustainability (ESG). ESG reporting has become a practice in larger corporate houses, in various formats.

Sustainability practices are not new to modern society—material and product recycling has a long history, and green construction has its roots in byproduct recycling. Before industry revolution, our common resources, by default, happened to be green. Steel and aluminum were recognized as strategic materials during World War II, and were recycled and reused to manufacture military equipment. Recycling programs fell into decline after the end of the war, but, since the oil shock in the 1970s, metals such as aluminum, copper and steel began to be recycled again because their production from ore is more energy intensive, and this soon became a preferred option (Roosa, 2010).

8.2 REGIONAL PERSPECTIVES ANALYSIS: COMPARISON TO GLOBAL AND REGIONAL (E.G. ASEAN) BEST PRACTICES

Iron and steel (I&S), at about 7% share of global GHG emissions, followed by aluminum at more than 1%, are the major energy consumers in the metal sector, which is energy intensive in nature.

Of the top ten I&S producing countries in 2017, contributing to 82.8% of global production, five were situated in Asia, three in Europe, one in North America and one in South America. Out of this 82.8% global production of I&S, 67% was from Asia, 9% from Europe, 4.8% from North America and 2% from South America. Similar regional distribution can be observed in case of the aluminum sector, as well (He et al., 2020).

A shift in production base of metal industries is also noticeable. For example, the total share of production of the top ten I&S producers in 2017 increased from little more than 64% in 1980 to 82.8% of global production. The respective changes on regional basis were from 21% in 1980 to 67% in 2017 for Asia, from 27% in 1980 to 9% in 2017 for Europe, from 14% in 1980 to 4.8% in 2017 for North America, while remaining at 2% during both years in South America.

Therefore, it has been observed that the production base of metal industries at large, and I&S in particular, has been shifting from Europe and North America to mainly Asia. While Asia in general—and Japan and South Korea in particular—have been successful to adopt the world's best practices of EE&C, countries in Europe in general—and Sweden in particular—have been ahead in achieving a 100% renewable energy (RE) path in the I&S industry.

A reason for this shifting of production may be the decrease in demand for production of metals from raw materials for developed countries because a large amount has already been produced and used for many years. Therefore, a large amount of scrap is available for recycling from metal products after the end of useful life. Metals can be recycled many times after the end of life of its products. As the production cost of metals from its scrap is only 5–15% of that of from its ore, the former would be the preferred manufacturing route to the latter the more a country accumulates the stock per capita of that metal products. In those countries and situations, the demand for metal products may increase, but not the production from its ore. A graphical representation of I&S stock in use (tons per capita) distribution all over world is shown in Figure 8.1 (Pauliuk et al., 2013). They reviewed the per capita in-use stocks of I&S in the earlier industrialized and developed countries, such as Germany, Japan, the United Kingdom and the United States, are between 11 and 16 tons, and the increasing pattern of stock accumulation has been slowing down or has come to a halt. This metal stock accumulation per capita has been growing quickly in the more recently industrialized and developed countries, such as South Korea and Portugal, and is currently in the range of 6–10 tons. They proposed the range of saturation for I&S to be a total of 11–15 tons, comprising 8–12 tons for construction, 0.8–1.8 tons for machinery, 0.8–2.2 tons for transportation, and 0.4–0.6 tons for appliances and containers. The global map of current stock (Figure 8.1) shows a regional distribution of I&S stock per capita and can give an idea of future metal market—particularly that of I&S—when compared with these saturation values.

FIGURE 8.1   Iron and steel stock in use (in tons per capita) all over world.

Source: Pauliuk et al. (2013)

One of the drivers of market is the demand for energy transition metals, such as for the advances of low-carbon energy technologies. The stress placed on people and the environment in metal ore extractive locations will be enhanced due to increased extraction rates. To quantify this stress, another regional effect of ESG risk of Energy Transition Metals (ETM) has been assessed by Éléonore Lèbre et al. (2020) and shown in Figure 8.2.

Some more regional and global perspectives in terms of environmental, social, economic and governance sustainability of metal industries has been discussed in subsequent sections of this chapter.

8.3   LESSONS FROM COVID-19

Metal industries suffered significantly due to the COVID-19 pandemic. The global average index of metals and mining industries has been calculated and graphically represented when compared to the Date of Outbreak (DO) of the pandemic. On DO, the index is taken as 100 points, for the purpose of showing its fluctuation during 30 days before DO and also 30 days after. Thus, the effect of the COVID-19 crisis has been compared with that of effect on index during other crises, such as the Global Financial Crisis, Ebola, H1N1 and MERS (Middle East respiratory syndrome) (Accenture, 2020). As is evident from Figure 8.3, COVID-19 had a significantly negative impact on the metal and mining industry index, second only to the negative impact of the Global Financial Crisis. Further study revealed that the negative impact originated from inactivity—particularly during lockdown periods—being declared periodically in almost every corner of the globe, and it has resulted in a number of effects, including the following.

- Reduction of market demand to a significantly high level due to huge inactivity in transportation, construction and many manufacturing plants.
- Supply chain disruption due to closings of borders and disruptions of raw material supplies from countries like China.

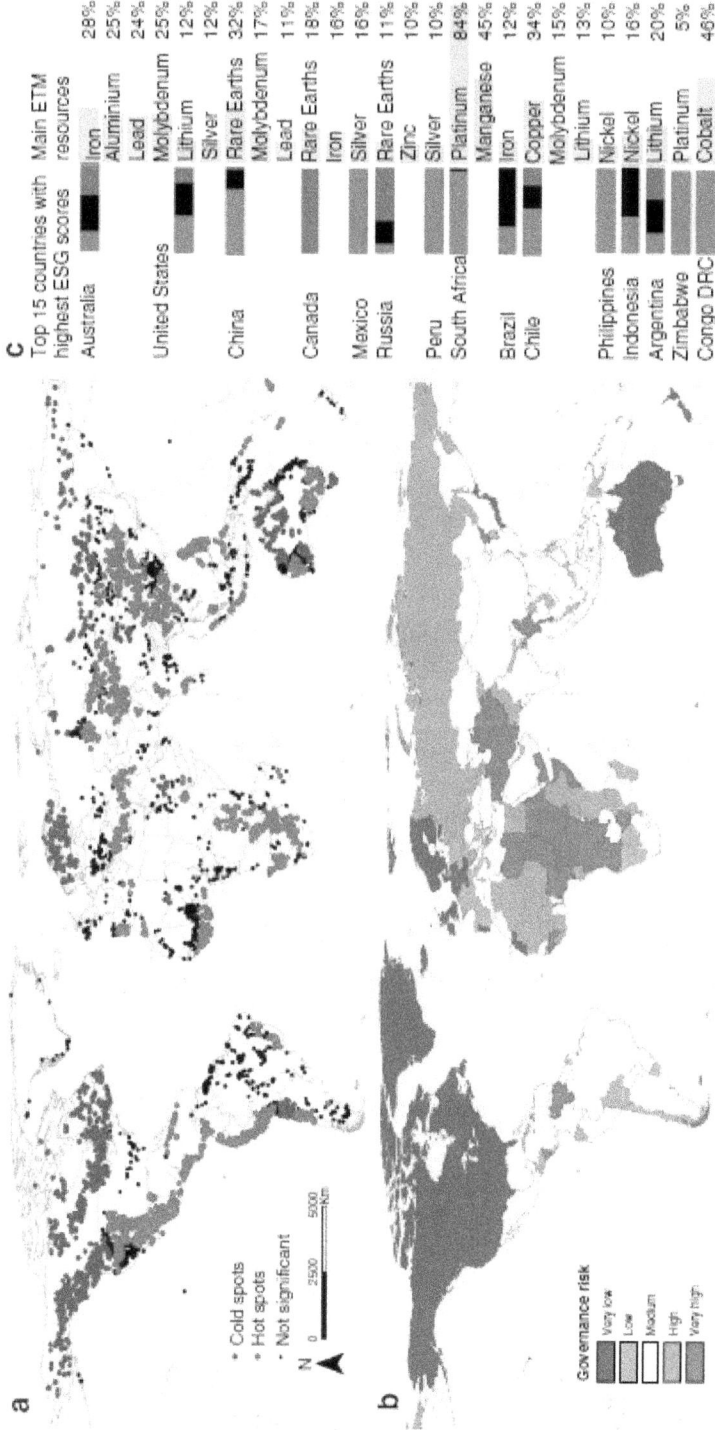

FIGURE 8.2 Global distribution of environmental social and governance (ESG) sustainability risks: a) social and environmental hot spots (clusters of significantly high environmental and social risk scores) and cold spots (clusters of significantly low environmental and social risk scores) are determined according to the sum of the six environmental and social dimensions, with mining projects outside hot spots and cold spots appearing as black dots; b) global governance risk map; c) top 15 countries according to the sum of individual mining projects' ESG scores and the percentage of main energy transition metal resources in each such country. Colored bars indicate the dominance of hot spots or cold spots in the country.

Source: Lèbre et al. (2020)

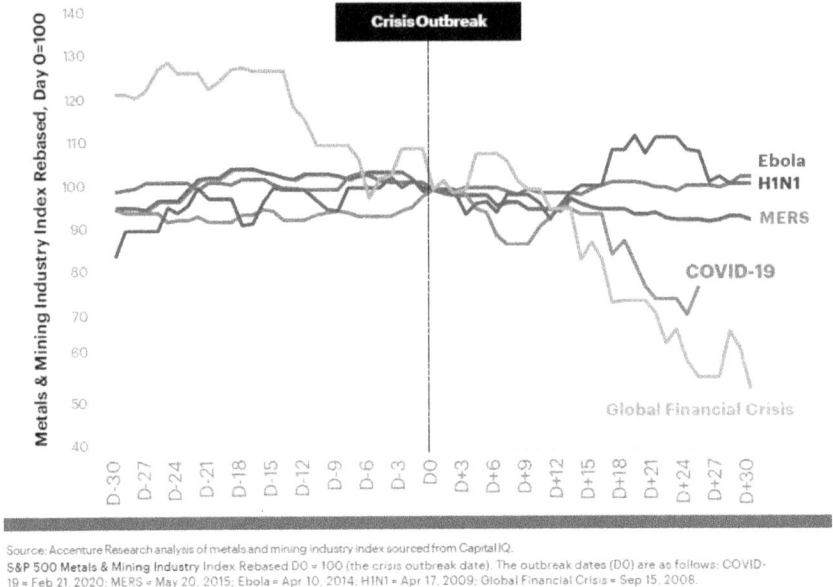

FIGURE 8.3 Impact of crisis on metals and mining industries due to COVID-19 during January 21–March 21, 2020 compared with other recent world crises.

Source: Accenture (2020)

- Supply of products notably hampered.
- Imposed lifestyle changes, cases of sickness and sometimes death has marred the spirit of the metal industry as a whole.

However, the demand will not be reduced, in fact may increase, in some of the emergency services related to metal industries value chains (Eurometaux, 2020) at times of pandemics as placed in the following table.

TABLE 8.1

Some of the metals and their products essential in the medical and food supply chains, infrastructure, etc.

Some Metals and Their Products	Essential Uses in Medical and Food Supply Chain
Aluminum profiles	Respiratory machines and other medical instruments
Aluminum can and foil sheet	Packaging of food, drinks, pharmaceutical and medical products
Copper rod	Ventilators and other electro-medical appliances
Zinc	Medical devices, ionization units, air purifiers, etc.
Stainless steel	Pharmaceutical processing and storage equipment for life-saving drugs, sterilization, medical instruments and equipment, etc.
Silver	Surgical tools, medical implants, electronic devices and for antimicrobial uses, etc.
Lead	Radiation protection in X-ray machines and laboratories

Platinum group metals	Pacemakers, chemotherapy drugs, brachytherapy, etc.
Some Metals and Their Products	**Essential Uses in Infrastructure and Power Supply**
Iron and steel	Essential and non-essential machinery and equipment, quick construction of hospitals, mask production facilities and infrastructure
Copper rod	Energy cables, electric motors and generators for energy transformation/distribution
Lead, nickel and other metals	Batteries for providing emergency power supply to hospitals
Zinc	Galvanizing steel in transportation, energy and public water infrastructure
Nickel in stainless steel	Water treatment and distribution systems

POINTS TO PONDER

Following are some metals and their products essential in maintaining power supplies.

- Many of the metal industries, such as iron, aluminum and zinc, are electrical energy–intensive and are able to provide demand response services to power grid operators. By planning their production, they help power grids to be operated at stable loads, as they can shift the electricity-intensive activities from peak load conditions to lean load conditions.
- As in most cases, metals smelters and refiners operate continuous production processes; their sudden shutdown incurs huge economic or physical damage to the tune of 200 million Euros and 400 million Euros to restart and production loss of several months for reconstruction [authors' analysis from HSE (2001) and Eurometal (2021)]. Metal industries also maintain captive power plants, which can also serve power grids at time of crisis.

8.3.1 TAKEAWAYS FROM COVID-19 LESSONS

- Maintaining facilities with safety issues in place with less staff at site, working from home and inventory management at high level of uncertainty are the challenges to be overcome in metal industry.
- High levels of cooperation across industries can minimize the impact of supply chain disruptions and falling demand.
- The IoT needs to be utilized in metal industries to help reduce the impact on operations and communication, even in case of such crises in the future.
- Innovation, teamwork and constructive thinking can help the turnaround of metal industries from a most challenging situation.
- The COVID-19 pandemic has given us an opportunity to understand our genuine needs vis-à-vis avoidable or non-productive consumption of resources. This would enable the general public at large and metal industries product designers and planners in particular to outlay action plans to adapt to a "new normal" after the crisis is overcome. Productive use of goods

and services, and 5S (sort, set in order, shine, standardize, sustain) and 5R (refuse, reduce, reuse, repurpose, recycle) practices in industry, may appear tough when taken up seriously, but would help directly and indirectly to attain the 17 SDGs by 2030 and restrict the Earth's temperature rise within 1.5°C by 2050 compared to pre-industry levels.

POINT TO PONDER

Maintaining and enhancing EE&C during any crisis situation is more difficult also because efficiency of any equipment and process tends to decrease at partial load. A thorough restructuring and renovating of the system would be necessary.

8.4 ECONOMIC, ENVIRONMENTAL AND SOCIAL SUSTAINABILITY

The corporate world also has begun to adopt policies for long-term sustainability of business. It identified three pillars—economic, environmental and social sustainability—by prioritizing some of the SDGs more relevant to industry. The response to SDGs by the corporate world in general is, however, mixed. To address the energy trilemma as mentioned in Chapters 1 and 2, RE and EE&C in metal industries would play equally important roles for meeting the objectives of equitable economic development, along with decarbonization and environmental sustainability. Now, industries are making more governance decisions, such as environmental and social factors, in the assessment of stock value, and thus, good governance includes the objective of economic sustainability.

Eight sustainability indicators, classified under three categories—environmental (e.g. CO_2 intensity, energy intensity, material efficiency and Environmental Management System [EMS] readiness), social (e.g. lost-time injury frequency rate and employee training) and economic (e.g. investment in new processes and products, and economic value distribution)—have been identified (World Steel Association, 2020) and collected for 87 I&S industries around the globe during the period from 2003–2019. The data has been analyzed for a decade (2010–2019); some of the observations have been placed in Table 8.1.

8.4.1 ADVANCES IN SUSTAINABILITY AND ESG REPORTING OF METAL INDUSTRIES

Presently, the term ESG (environmental, social and corporate governance) is often used in the corporate world to address the issues of economic, environmental and social sustainability. The United Nations Conference on Trade and Development (UNCTAD), which has been working in the area of sustainability reporting for more than two decades, initially identified the following four core indicators for the SDGs in company reporting (UNCTAD, 2016, 2017): economic, environmental, social and institutional—with "institutional" subsequently replaced by governance in 2016 and 2017, respectively. The United Nations Global Compact (UNGC) is a nonbinding pact to encourage businesses worldwide adopting sustainable and socially responsible policies, through two distinct engagement levels to suit each company's

TABLE 8.1

Sustainability Indicators during 2003–2019 for 87 I&S Companies: Relevance and Change

Environmental Performance and Relevance											% Increase
2003	**2010**	**2011**	**2012**	**2013**	**2014**	**2015**	**2016**	**2017**	**2018**	**2019**	**

1. CO_2 intensity (tons CO_2 / ton crude steel cast): can decrease with increase of EE&C, recycling and reduced use of blast furnaces and coal. An increase of 1.7% is due to increase of production in Asia, dominated with primary route path with compared to Europe and the United States—some reduction due to improvement in EE&C is also offset during the decade.

1.80	1.76	1.75	1.82	1.80	1.87	1.87	1.84	1.81	1.83	1.7%

2. Energy intensity (GJ/ton crude steel cast): can decrease with increase of EE&C, recycling and reduced use of blast furnaces and coal. A decrease of 1.4% indicates that in spite of a shift of production share to Asia—and hence more use of the blast furnace route in the production of I&S—improvement in EE&C and recycling has offset the previous effect in favor of the latter.

20.1	19.8	19.6	20.1	19.8	20.3	20.3	19.9	19.5	19.8	−1.4%

3. Material efficiency (percentage of solid materials converted to products and co-products): this can be improved further by approaching zero waste practices; say, through 5S and 5R. Though there was significant improvement in the previous decade, no further improvement has been found in the most recent decade. This may, however, indicate that China and India are fast catching up with the waste minimization techniques, as the overall value has not changed even though production increments have taken place in these two countries when compared to the rest of the globe in the most recent decade.

96.1	97.5	96.1	96.5	98.0	97.5	97.4	96.9	96.5	96.3	97.5	0.0%

4. Environmental management system (EMS) readiness (percentage of employees and contractors working in EMS-registered production facilities): as envisioned throughout this book, and particularly in this chapter, the improvement in environmental awareness is significant in metal industries around the globe.

90.9	87.6	89.9	89.5	90.2	94.1	93.6	96.9	96.6	97.1	97.2	9.6%

Social Performance and Relevance

5. Lost-time injury frequency rate (injuries per million hours worked): globally, the awareness on safety has been increasing. China and India are fast catching up to the safety standards and best practices of Japan, Europe and the United States, leading to a global improvement of as much as 63.8%.

2.29	1.91	1.45	1.60	1.39	1.17	1.01	0.97	0.84	0.83	−63.8%

6. Employee training (training days per employee): maybe because of more automation in the new production systems, training days per employee has rather decreased a bit. Here is a scope to improve through more training, skill development and awareness programs on EE&C.

7.5	7.0	7.7	7.9	7.8	6.3	6.8	6.9	6.3	6.4	6.9	−0.9%

Economic Performance and Relevance

7. Investment in new processes and products (percentage of revenue): now it is time for the industry to go for greener and cleaner production of metals by increasing the investment to these new areas for improving sustainability, instead of lowering it by 1.7%.

6.37	8.80	8.28	10.05	8.53	7.32	8.22	7.70	5.76	6.10	7.07	−1.7%

(Continued)

TABLE 8.1 Continued

Economic Performance and Relevance											% Increase
2003	2010	2011	2012	2013	2014	2015	2016	2017	2018	2019	**

8. Economic value distributed (percentage of revenue): Thanks to larger CSR initiatives around the globe, an improvement has been taking place in the recent decade by 4.6%.

2003	2010	2011	2012	2013	2014	2015	2016	2017	2018	2019	% Increase
	93.5	95.7	99.8	96.8	96.3	100.1	97.5	95.4	93.8	98.0	4.6%

** when compared to 2010 (one decade)

Notes:

Indicators 1 and 2: CO_2 intensity and energy intensity are calculated using route-specific energy and CO_2 intensity for the BOF and EAF. The indicators are also weighted based on the production share of each route. Data prior to 2007 is not available.

Indicator 5: Lost-time injury frequency rate includes fatalities and is calculated based on figures including contractors and employees. Data prior to 2004 is not available.

Indicator 6: Employee training includes production and non-production facilities.

Indicator 7: Investment in new processes and products includes capital expenditure and R&D investment.

Indicator 8: Data collection for EVD started in 2007.

Source: World Steel Association (2020).

needs: participant or signatory (UNGLOBALCOMPACT, 2021). Similarly, Global Reporting Initiative (GRI, 2022), an independent, international organization, helps businesses and other organizations take responsibility for their impacts by providing them with the GRI Standards—the world's most widely used standards for sustainability reporting. Other popular reporting platforms are provided by ISO (2022), CFI (2021) and CDP (2021). ISO, the International Organization for Standardization, develops and publishes international standards on various fronts and tries to promote sustainability. ISO 26000 in particular provides guidance to corporations which recognize that respect for society and the environment is a critical success factor and provides them a way of assessing their overall performance and declaring their commitment to sustainability. Other relevant ISO standards include the following.

- ISO 14025 (2006): environmental labels and declarations.
- ISO 14040 (2006): environmental management—life cycle assessment—principles and framework.
- ISO 14044 (2006): environmental management—life cycle assessment—requirements and guidelines.
- ISO 21930 (2007): sustainability in building construction (currently being updated).
- ISO TS 14067 (2013): carbon footprints (technical specifications).
- ISO 14046 (2014) water footprints.

CDP (formerly known as Carbon Disclosure Projects) is a 20-year-old not-for-profit charitable organization known worldwide for its contribution toward establishment and effective functioning of the global disclosure system for investors, companies, cities, states and regions to manage their environmental impacts, using various indigenously developed instruments including CDP Score.

Another relevant tool is the GHG Reporting Protocol outlined by WRI/WBCSD.

In 2018, UNCTAD found out the following for 100 global top listed companies, as well as the following ESG rating agencies' use of indicators (UNCTAD, 2018).

- 99% produce some sort of ESG reporting.
- 85% produce stand-alone corporate reporting; of this, nine produce integrated/combined reports.
- On an average, ESG reports are 100 pages each in length, with information on 49 indicators.
- While 42% did not define any boundaries, 5% used financial boundaries, 9% used operational boundaries and the rest (42%) used homemade/conventional boundaries.
- 40% provided their ESG accounted principles in detail, 12% provided them in part, and the rest (48%) did not mention their accounting principles.
- 72% of corporates refer to GRI standards, 62% to CDP standards, 58% to ISO standards and 51% to UNGC standards. It is evident that some corporates are conforming to more than one method of reporting.
- 57% got some external review by renowned firms.
- Energy use is reported by 78% of the secondary, 60% of the conglomerate, 47% of tertiary and 20% of the primary sector corporates; overall, this is 57%, as shown in the Table 8.2. Therefore, EE&C has further to go for all industries to attain sustainability.

Categories	Indicators per category	Frequency
Environmental	CO_2e (scope 1 + 2)[35]	82[36]
	Water consumption	74
	Waste	63
	Energy	57
	Reuse of waste	51
Social	Number of employees	93
	Donations (incl. community projects and employees' voluntary work, etc.)	89[37]
	Number of employees divided by gender	66
	Number of managers divided by gender	62
Governance	Number of female board members	99
	Existence of Audit Committee	97
	Compensation – total	95
	Attendance rate to board meetings	87
	Compensation per member	86
	Board duration of service (tenure)	84
	Number of board meetings	83
	Age diversity of board members	77
	Number of Audit Committee meetings	74
	SOX activities	74
	Existence of ESG Board, CSR Committee, Corporate Governance committee, etc.	50

TABLE 8.2   Most-Used Sustainability Indicators by the Corporates

Source: UNCTAD (2018)

- 92% disagreed that current reporting is sufficiently comparable. This also indicates that the awareness and popularity of ESG reporting is increasing.
- 12% are using metal a major input in the production process, such as metal manufacturing and mining, auto and truck manufacturers, etc.

ESG sustainability reporting by metal industries will gain increased attention of society in view of the increasing pressure on Earth's resources for two primary reasons: first, increasing population at least until 2050, and second, increasing standard of living, as well as demand for more and more value added services.

POINT TO PONDER

Among all industry sectors, the metal and mining industries assumed the highest ESG sustainability risk (scored 11 in a scale of 1–12), according to S&P Global (2020), and hence need concerted attention to achieve sustainability.

8.5 CIRCULAR ECONOMY

A circular economy refers to a move from linear business models in which products are manufactured from raw materials and then discarded, to circular business models where products or parts are repaired, reused, returned or recycled (World Economic Forum, 2021).

Eurometaux (2015) has pointed out that non-ferrous metals, because of their longer recyclability when compared with ferrous metals, would attain central stage in 2050's sustainable society. In the process of attaining circular economy, non-metals will become integral part of the society and will facilitate innovation, deliver mobility, enhance communications, protect products through packaging and lower energy consumption in many sectors.

Metal industries must take practical actions in order to achieve circular economy in reality through business leadership, international cooperation and sharing of best practices among the partners in the value chain and smart regulatory framework. The industry should broaden the skill sets of their employees and different skill sectors according to employability and national skill quality framework applicable for different countries, building constructive relationships across all stakeholders and accelerating among all concerned adoption of the idea of reduce, reuse, remanufacture and recycle as depicted in Figure 8.4. The metal industries also need to be proactive in strategic planning to face challenges on least in four fronts: demographics (and population explosion), stability in energy and climate, raw materials supply and environmental degradation.

Implementation of circular economy therefore calls for several initiatives, as prevention is better than a belated cure, including the following.

- To accelerate and replicate the practice of secondary metals recycling, which helps to significantly reduce energy and raw material input and facilitates the development and functioning of secondary metal markets: metals, such as steel, have much higher recyclability when compared to wood, concrete or plastic.

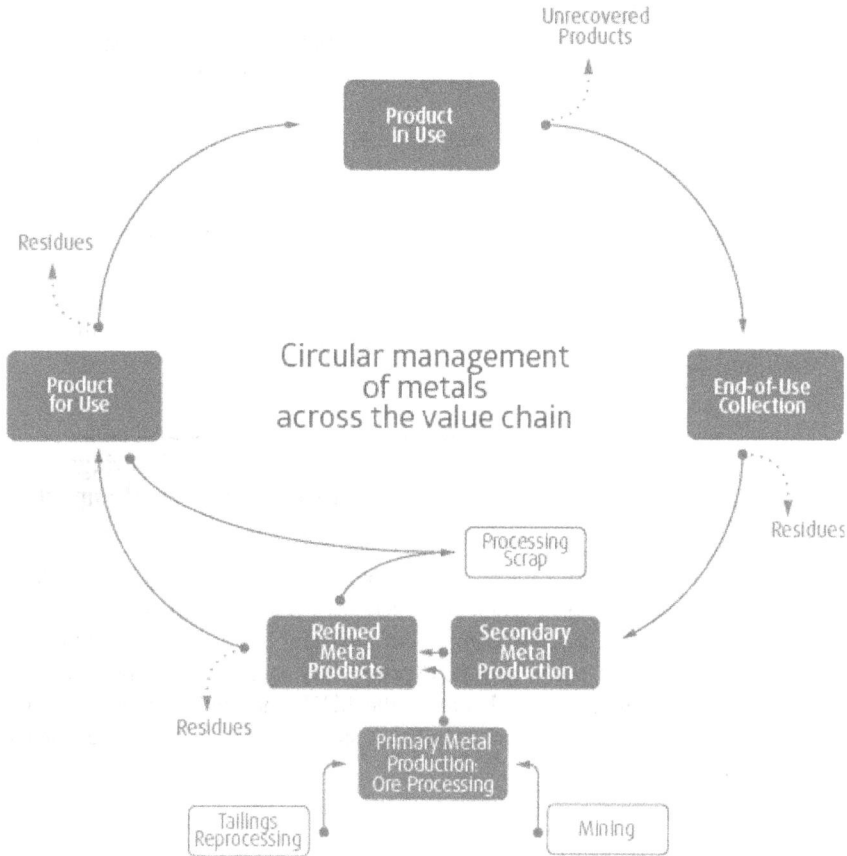

FIGURE 8.4 Circular economy in metal industries through circular management across the value chain.

Source: Eurometaux (2015).

- To encourage making it compulsory at every base metals production plant to establish circular economy in order to link the productive utilization of byproducts and recycled metals with the aim of increasing their profitability. The large potential of product and material substitution and increasing the use of scrap metal and large amount of waste in steel and aluminum production must be emphasized. Metal structures can be designed to be much lighter than wood, concrete or plastic.
- A life cycle assessment (LCA) tool can be used to measure the holistic economic and/or environmental impact or performance of a product at each stage in its life cycle, "from cradle to grave." LCA then helps to compare the economic and/or environmental sustainability of similar products and services which have the same function/purpose to an individual or society.

- To establish a market-driven mechanism for recycling materials supported by inter alia green certificates for recycled materials, waste legislation (at all walks of life, including producers' responsibilities), mandates for eco-design and fiscal incentives.
- To ensure cohesion of policy and funding, so as to promote resource efficiency and recycling.
- To set collection targets, stress the need for qualified and skilled people to cope with the transition toward more sustainable production processes and products.
- To support companies, research institutes and the social partners to jointly investigate skills needs for ESG sustainability and accordingly define the training and education strategy.

Case Study 8.1: Tata Steel Europe—Circular Economy in Bridge Construction Found Steel to Have the Lowest Environmental Impact

Application of circular economy concepts as mentioned in this chapter has become useful to identify a more sustainable path in infrastructure development. An independent research on life cycle assessment (LCA) on bridges of concrete, concrete with asphalt, composites and steel has been reported (World Steel Association, 2015). The results, expressed as an environmental indicator (MKI) score (in Euros), showed that as the MKI, the environmental impact of the project also increases, and found that steel has lowest life cycle impact (Figure 8.5).

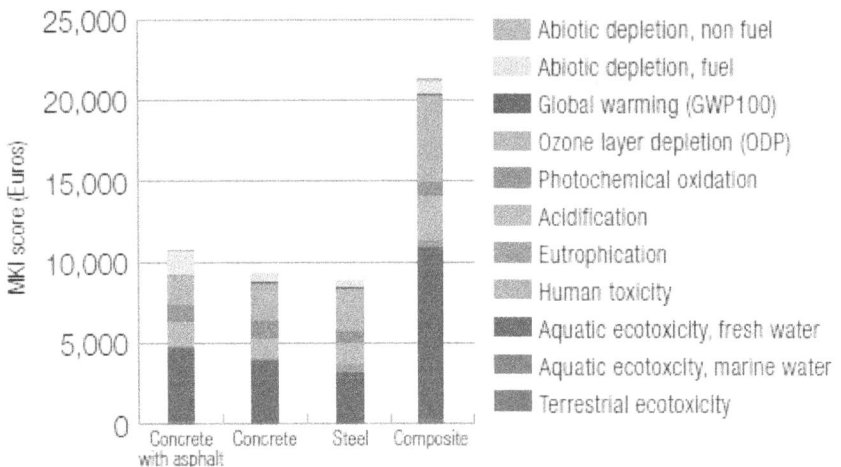

FIGURE 8.5   MKI scores for road bridges shows steel has the lowest impact over the life cycle of the bridge.

Source: World Steel Association (2015).

Case Study 8.2: China Steel Corporation—Circular Economy Applied to Small Motors Reveals That Steel Has Scope for Continuous Development

Continuous development is essential condition for sustainability. For example, advanced electrical steels (50CS290 and 50CS400) when compared to the least advanced steel (50CS1300), when used in 3-horsepower electric motors (Worldsteel, 2015), demonstrated significant (2.9%) improvements in motor efficiency, reduced greenhouse gas emissions over the life cycle and increases in shaft lifetime to 20,230 hours, an increase of 24.5%. This is also an example of cooperation among four stakeholders: China Steel Corporation (CSC), the Tatung Motor Co., the Industrial Technology Research Institute (ITRI) and the Metal Industries Research Development Centre (MIRDC). The findings are shown in Figure 8.6.

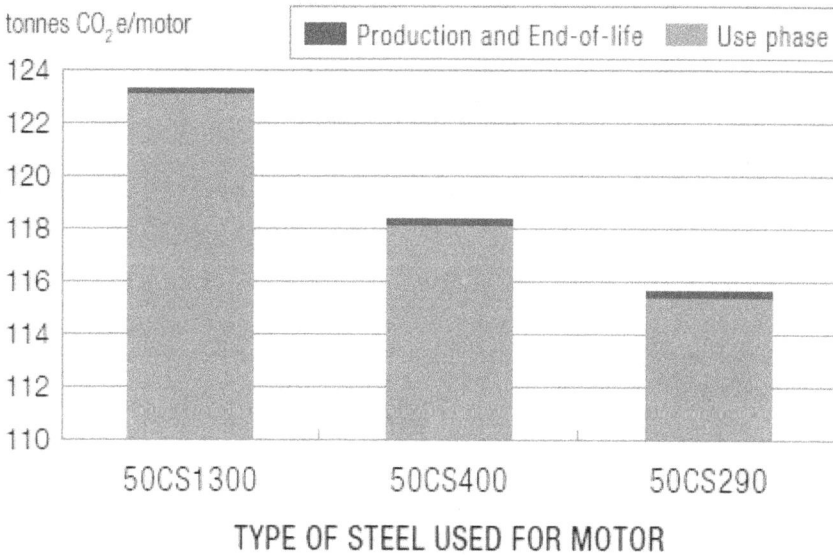

FIGURE 8.6 Life cycle emission in tons CO_2e/motor for various types of steel used for small AC motors.

Source: World Steel Association (2015).

TIPS ON TOP

Most base metals can be said to have infinite life, as they can be recycled many times. The energy required to produce most base metals from their recycled scrap is only a fraction of the energy required to produce it from ore.

If lack of policy or technology is a cause of long-distance trade, change the policy and/or innovate/invest for technology to reduce it as long distance trade, particularly that of metals, in the long run hindering sustainability.

8.6   EXTENDED PRODUCER RESPONSIBILITY

We do not inherit the Earth from our ancestors; we borrow it from our children.

– Native American Proverb

Resources are collective property of not only the present generation, but also that of future generations. Therefore, the metal upstream-mainstream-downstream industries share the responsibility of non-destructive use of metal resources. It is possible to achieve such objective through extended producer responsibility.

Society as a whole will approach 100% recycling of metals after the normal useful life of the product is over, leading to almost zero requirements for mining and exploration for ores (Eurometaux, 2015). For example, the earlier proposed range of saturation for I&S to be a total of 11–15 tons per capita could be significantly lowered, particularly for the presently developing countries, through collaborative engagement of stakeholders in all economic activities including resource efficiency and innovation in a reshaped industrial ecosystem with government and communities. The European Union envisions that in 2050, industry will enjoy a relation of trust and respect with other partners to respond to 21st century environmental challenges as an integral partner in society (Eurometaux, 2015).

As most metals have almost infinite life, 100% recycling of products would be feasible and possible through practice of reverse logistics, segregation at source, refuse-reduce-reuse-repurpose-recycle (5R), circular economy, lean manufacturing, LCA of products, Kaizen (5S: sort, set in order, shine, standardize and sustain), zero waste, zero defect, Race to Zero campaign, and so on. These concepts developed in modern times and as successor of traditional oriental value system the East: "Earth is my Mother and I am her child."

Kaizen—the idea first developed in Japan—strives to establish good working environment by reducing "Muri" (i.e. excessiveness), "Mura" (i.e. inconsistency—material or spiritual) and "Muda" (i.e. wastefulness) to restore appropriateness of quantity, integrity of quality and productive activities, respectively (Figure 8.7 and Figure 8.8).

POINT TO PONDER

Even 100% recycling of metals through extended producer responsibility and other necessary mechanisms would not be enough to halt exploration or production of base metals through primary route because of growing population, increasing standards of living and switching over to renewables. This is because new installation of wind turbines and solar technologies—for the same installed power generation capability—consume up to 90 times more aluminum; 50 times more iron, copper and glass; and 15 times more concrete, besides rare and critical metals, than the conventional thermal (oil, natural gas or nuclear) power stations currently operating (European Parliament, 2015). Metals used for such alternative modern energy systems are collectively known as energy transition metals (ETMs).

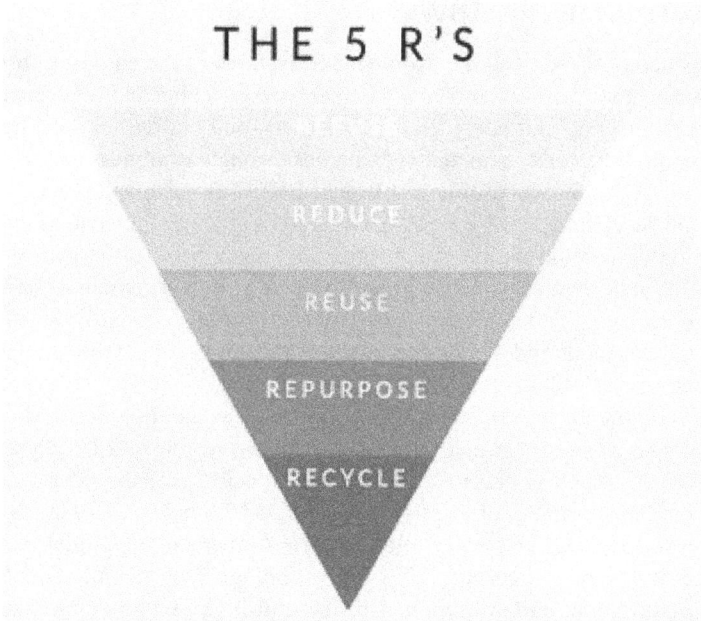

FIGURE 8.7 Metal life could be enhanced through the 5R practices.

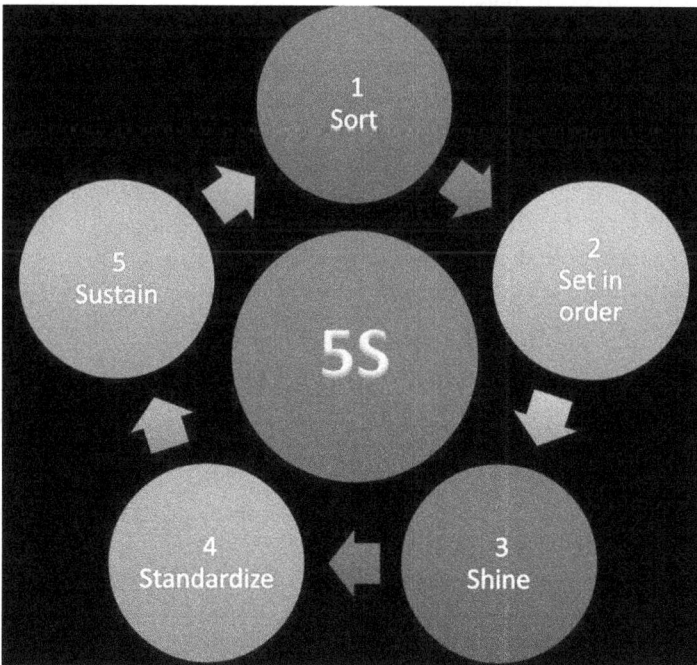

FIGURE 8.8 Metal waste could be reduced through the 5S practices.

8.7   DEVELOPMENT PATHWAY

A convergence of top-down—i.e. from United Nations to the end user, through the INDC, energy policy, EE&C projects/schemes, energy audits and implementation of its recommendations, ultimately user satisfaction—and bottom-up—i.e. from users and workers to the board room and apex bodies through awareness and small group activities [SGA], recognition of demonstrated EE&C projects and practices, acceptance at top level followed by replication and adoption at wider [regional/global] scale—initiatives would be the watchwords for a sustainable development pathway for the metal industries. This calls for parallel action on many fronts—international, national, corporate, academics, non-governmental organizations (NGOs)—in conceiving future agendas and the implementation and monitoring of the same for ensuring success.

In line with the Paris Agreement objective to pursue an all-out effort to keep the global temperature rise preferably to 1.5°C, in any case not beyond 2°C, the European Commission proposed to raise their target on reducing greenhouse gas emissions from the previous target of at least 40% to at least 55% below 1990 levels by 2030 (European Union, 2021). Re-entry into the Paris Agreement on Climate Change by the United States on 21 February 2021 brings the figure of member countries producing half of global carbon pollution to two-thirds. In a continuously degrading situation, the role of metal industries is important in view of global efforts toward net zero, with ambitious 2030 and 2050/2060 targets. and taking more steps in advance of COP26 in Glasgow in latter half of 2021 (UNFCCC, 2021a).

Besides the EE&C technologies and techniques, many of which have been discussed throughout this book, some innovative approaches are unique in their own right—amazing amidst our plight—and placed only briefly to be in our sight.

1. Michel Y. Haller (2020) shows that 7–11 kWp of photovoltaic installations and 350–530 kg of aluminum would be needed per apartment for different Swiss climates to meet all the needs at home, taking care of poor availability of sunshine in winter. Aluminum is used as seasonal storage to meet all the needs in winter from the surplus solar energy stored in summer. It is evident from the main working principle, as shown in Figure 8.9 and Figure 8.10, that this approach fulfills multiple objectives, including resource conservation, higher efficiency, reduced life cycle cost, higher productivity and an improved environment.

2. Carbon capture and storage/utilize (CCS/CCU) technology captures CO_2 emissions at their source and stores them in a safe long-term underground storage facility or processes it for productive use. Though theoretically the idea seems to be ideal for carbon-intensive industries such as metal industries, particularly aluminum and I&S, to date there is no large-scale commercially successful project in the metal sector, except the one at the Emirates Steel Industries (ESI) CCS Project—the world's first iron and steel project to apply CCS at large scale. Since 2016, it captures around 0.8 Mt/y of CO_2 from gases produced by the DRI reactor, according to Global CCS Institute (Global CCS Institute, 2017).

TABLE 8.3

Steel Making Energy Use (GJ/Metric Ton) and Ultimate Saving Potential

Process	Absolute Minimum	Practical Minimum	Actual Average Requirement	% Over Practical Minimum
Liquid metal "pig iron"	9.8	10.4	13.5	23%
Liquid hot metal: electric arc furnace	1.3	1.6	2.25	29%
Liquid hot metal: basic oxygen furnace	7.9	8.2	11	25%
Hot rolling flat	0.03	0.9	2.2	59%
Cold rolling flat	0.02	0.02	1.2	98%

Source: http://large.stanford.edu/courses/2016/ph240/martelaro1/

3. According to theory, electrolysis of aluminum starting from bauxite should consume only 6.23 kWh/kg of Al at highest efficiency point, including the energy loss due to consumption of carbon electrode being consumed into carbon dioxide emissions (CO_2). Unfortunately, aluminum (Al) is one of the most carbon-intensive, electricity-intensive and energy-intensive materials to produce—highest on a per-weight basis. However, the aluminum industry has significant opportunities to reduce these intensities. For example, the most energy-intensive process in aluminum industry, the smelting process in Hall–Héroult technology, requires 15.37 kWh/kg of Al, while the theoretical minimum energy requirement for this process is 6.23 kWh/kg of Al. Similar to Carnot cycle efficiency in heat engines, this theoretical limit is a source of motivation for the researchers and operators to strive for further reduction until the theoretical limit is achieved (Obaidat et al., 2018).

4. Similarly, the theoretical limit for energy consumption for production of I&S from ore (hematite) is much less than what is actual average, as shown in Table 8.3. The potential, however, is not so high if compared with the reduction already achieved—almost 200 GJ/t of pig iron in 1800 to less than 100 GJ/t by 1850, to only about 50 GJ/t by 1900, and to less than 20 GJ/t in 2000s. Still, the efforts to approach the theoretical limit is considered a cost-cutting and emissions control measure.

The following case studies show that coordinated and determined effort could halt the degradation of nature due to anthropogenic intervention.

Case Study 8.3: Selected Advanced Companies in Europe With the Target of an Expected Reduction of Specific CO_2 Emissions of 50%

Ultra low–carbon dioxide steelmaking (ULCOS) is an initiative launched by the major players in the European steel industry and its main partners in other industries and academia (47 partners, 15 European countries) in the context of the post-Kyoto era. With the target of an expected reduction of specific CO_2 emissions

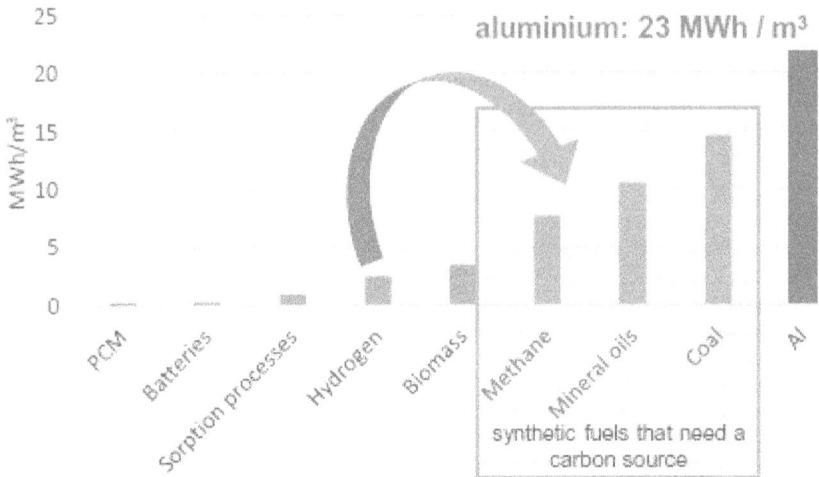

FIGURE 8.9 Energy densities of different storage technologies on a volume basis.

Note: PCM = phase change materials

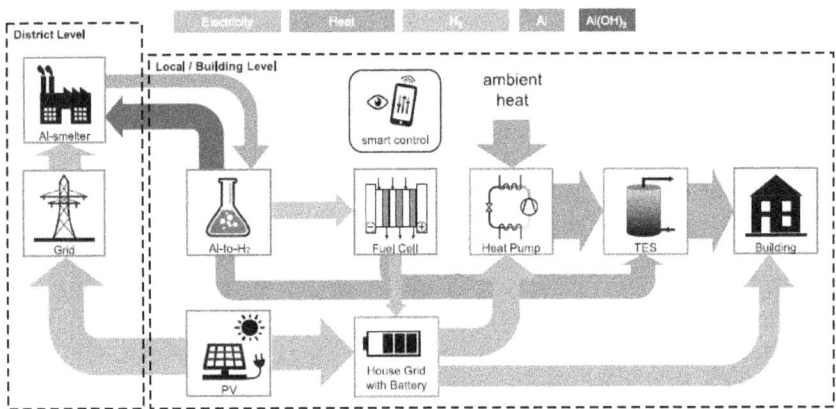

FIGURE 8.10 Schematic system concept for heat and electricity provided 100% by a photovoltaic and heat pump system, with a seasonal aluminum redox storage cycle.

of 50% as compared to a modern blast furnace, the following seven panels of technologies have been examined in the first stage of the proposed stage-gate approach: (i) new carbon-based smelting-reduction concepts, making use of the shaft furnace but also of (ii) new less common reactors; (iii) natural gas–based pre-reduction reactors beyond state-of-the-art technology; (iv) hydrogen-based reduction using hydrogen from CO_2-lean technologies; (v) direct production of steel by electrolysis; (vi) the use of biomass, which circulates carbon rapidly in the atmosphere; and (vii) CO_2 capture and storage will be included in the design (https://cordis.europa.eu/project/id/515960).

Europe's largest iron ore producer, LKAB of Sweden, plans to invest almost 40 billion Euros over the next two decades in emissions-free steel production. LKAB, along with Vattenfall and SSAB, are behind the HYBRIT project which intends to grow Sweden's steel industry, fossil free. They will use hydrogen instead of coal as the "reducing agent" to remove the oxygen from the iron ore (Thomas Koch Blank, 2020). There are at least three important aspects for which this initiative is a significant landmark in the world of steel:

First, implementation of this initiative of LKAB, by avoiding the use of blast furnace, will reduce more than 50% GHG emission of Sweden's total footprint. It may be noted that many of the blast furnaces are located in other nations.

Second, the hydrogen produced through renewable routes would significantly contribute to lower the cost of this zero-carbon fuel. This initiative would also help the other sectors of economy, such as aviation and shipping, to address emissions.

Third, the results of the ongoing process trials indicated that this technology will be commercially scalable.

Source: European Union (2010).

Case Study 8.4: Race to Zero

Race to Zero is a global campaign of 120 countries in the largest ever alliance (called Climate Ambition Alliance, which was launched at the UN Secretary-General's Climate Action Summit 2019 hosted by Chile President Sebastián Piñera), outside of national governments, committed to achieving net zero carbon emissions by 2050 at the latest. It calls for leadership and support from businesses, cities, regions and investors for a healthy, resilient, zero-carbon recovery that prevents future threats, creates decent jobs and unlocks inclusive, sustainable growth.

Already, Race to Zero mobilized a coalition of leading net zero initiatives, representing 471 cities, 23 regions, 1,675 businesses, 85 of the biggest investors and 569 universities as "real economy" actors, now covering nearly 25% of global CO_2 emissions and over 50% of GDP. The objective is to build momentum around the shift to a decarbonized economy ahead of COP26, and send governments a resounding signal that business, cities, regions and investors are united in meeting the Paris Agreement goals and creating a more inclusive and resilient economy (UNFCCC, 2021b).

Case Study 8.5: Tata Steel Ltd. India

We believe that Tata Steel has comprehensive monitoring of resource usage and emissions from its manufacturing processes. In addition to necessary permits being in place and compliance with regulations, the company has active programs to reduce these impacts over time, which in our opinion is commensurate with the significant environmental impact of the steel making process. In our opinion, the periodic renewal or continuity of local resource use, land use and pollution permit requirements at Tata Steel's India iron ore mining operations is an ongoing business risk that sets the company apart from its global peers. Tata Steel's labor, vendor and community relationship management initiatives are on par with large metals and mining companies we rate.

Going forward, we expect no significant financial contribution from Tata Steel for the funding of pension plan at its European operations. We believe that Tata Steel demonstrates satisfactory governance standards led by an independent board that is an active decision-making body in areas such as business ethics, management remuneration, and grievance redressal. There are no significant regulatory, tax or legal liabilities, and we believe there is transparency in financial reporting as well as consistency in communication of messages. The company secured from S&P Global Issuer Credit Rating: BB-/ Comments: Positive/--

Source: S&P Global (2019).

Case Study 8.6: HIsarna Innovative Steel making Process Developed by Tata Steel Europe

As part of the international ULCOS program, HIsarna is a smelting-reduction process in which iron ore is directly converted into liquid iron. This technology does not require the preparation of iron ore agglomerates or the production of coke. Without these preparatory steps, the HIsarna process can use the raw materials more economically. It also requires less energy to operate. Compared to blast furnace steel production, it can reduce CO_2 emissions by 20%. This technology produces flue gas with very high CO_2 concentration (above 90%) because pure oxygen is used instead of air to feed the reaction. It does not require CO_2 separation beyond simple water removal. This can lead to significant savings in capital and operating costs associated to CO_2 capture. Therefore it is extremely attractive for CCS equipped steel plants of the future. Pilot-scale testing of the HIsarna process started in 2011 at Tata Steel's site in Ijmuiden, the Netherlands. Since 2011, there have been four test campaigns. The fifth campaign, scheduled to start in 2017, was to be a so called an "endurance test" lasting about 3–6 months to test long-term operational and maintenance aspects of the process and equipment. Achievement of these objectives through the 19 days longest run trajectory – producing 2,000 tonnes of pig iron proved that HIsarna is a very promising technology that could be soon deployed as substitute to BF/BOF process after demonstration at commercial scale will be executed. Subsequent deployment of second target test plant and long term ambitious policy aims to have a HIsarna factory on an industrial scale by 2030.

As the first steel manufacturer to become an approved Environmental Product Declaration (EPD) program operator, Tata Steel now have the ability to create product specific EPDs that comply with EN 15804 and ISO 14025 standards.

To meet the ambition of carbon neutrality by 2050, Tata Steel continues to develop and invest in a mix of technologies that will help to significantly reduce emissions over time. One such example is HIsarna technology, developed as an alternative to the blast furnace process to make the steel of the future with a minimum 20% reduction in CO_2 emissions. If testing on an industrial scale proves to be a success, this technology could come to market in 5–10 years' time.

At that point, Tata Steel could recapture zinc from recycled scrap steel, achieve at least 20% CO_2 reduction and—if combined with CCUS—achieve a reduction in CO_2 emissions of up to 80%. An additional advantage will be the elimination of emissions of nitrogen (Nox), decarbo oxides and particulate matter.

Sources: Global CCS Institute (2017) and Tata Steel (2020).

Case Study 8.7: Neves Corvo Mine Case Study

Stretching over 18% of the EU's land area and more than 8% of its marine territory, Natura 2000 is the largest coordinated network of protected areas in the world. It offers a haven to Europe's most valuable and threatened species and habitats.

The Neves Corvo mine is one of EU's largest underground copper mines. The mining concession area (1.619 ha) and the industrial area (840 ha) partially overlap with two Natura 2000 sites: the SPA Castro Verde (PTZPE0046) on the north of the industrial area, and the Guadiana SCI (PTCON0036). The totality of the concession's non-mining areas are managed in order to maintain biodiversity. As part of the EIA procedure, several potential environmental risks were identified, which posed possible risks some habitats and species included in the Natura 2000 sites, in case of an accident or inappropriate environmental management. In particular, there are risks from the accidental rupture or overflow from the tailing (sedimentation) ponds to the Oeiras River that passes through the mining area. Taking into account the EIA, Somincor set up a very strict EMS that includes a permanent and comprehensive soil and water monitoring with the aim of minimizing and mitigating any unexpected impacts, but also to identify opportunities for habitat and species' management. Somincor created partnerships with nature conservation NGOs, local farmers and local and national authorities to promote biodiversity conservation, in particular in Natura sites management and habitat restoration.

Source: European Union (2019).

TABLE 8.4

Holistic Approach for Sustainability in Metal Industries and Some Necessary Conditions

Sr No.	Use of Metals to Create More Sustainable Products, Services and Infrastructure in a Sustainable Market of At Least a Few Centuries Ahead	Conditions of Sustainability in Metal Industries Having a Sustainable Market
1	Renewable energy sources: metals are essential to manufacture any renewable energy systems. For example, a modern wind turbine system needs metals for building about 90% of its weight. Fuel cells need platinum; energy storage systems need cobalt, aluminum, lead, lithium, nickel and copper. Photovoltaic cells, solar thermal systems and hydroelectricity need metals such as nickel, zinc, copper and potassium.	**Interdisciplinary**: Chapter 2, while giving an overview on EE&C in major metal industries, emphasized the findings of Oliver Wyman and the World Energy Council on the energy trilemma: affordability and access, energy security and environmental sustainability (World Energy Council, 2021), as metal industries are consuming energy and emitting CO_2 to the maximum in the industry sector.
2	Low-carbon transport: because of their light weight, metals such as aluminum and magnesium are essential components in alloys used in vehicles particularly for reduction of weight, such as in automobiles and airplanes. Precious metals such as platinum are also critical to catalytic converters.	**Policy**: in Chapter 2, it was shown how holistic and sectoral policy can effectively help overcome visible and invisible barriers in various countries which again have adopted policies with focus gradually shifting from market forces (until 1973) to energy security and economic competitiveness (during 1973–1982) and since 1982 toward climate change, always giving high importance to the metal sector. Sustainability of metal industries has already drawn very high attention in the EU and elsewhere (Eurometaux, 2015). **EE&C technologies**: Chapter 3 discussed more important EE&C technologies which are pillars for sustainability in metal industries. Sengupta et al. (2018) identify, for example, four technologies only in WHR in metal industries, from across the global practices.

(Continued)

TABLE 8.4 Continued

Sr No.	Use of Metals to Create More Sustainable Products, Services and Infrastructure in a Sustainable Market of At Least a Few Centuries Ahead	Conditions of Sustainability in Metal Industries Having a Sustainable Market
3	Sustainable buildings: steel, aluminum, copper, lead, nickel and zinc have been used extensively in modern safe and sustainable buildings and construction for recyclability, durability (last 100 years or more without maintenance) and strength, resistance to corrosion or even for semiconductor property of silicon being used extensively in electronic gadgets and power control systems in buildings.	**Manufacturing processes:** as social cost is gradually becoming inclusive through various policy initiatives and climate concern at global level, clean, green, lean metal manufacturing processes are fast occupying center stage, along with productivity rise through EE&C technologies in place. Chapter 4 showcases a few cases demonstrating huge potential for sustainability of these industries taking the sustainable path of development.
4	Efficient communications: metals such as nickel, cobalt, copper and lithium are essential nowadays, being used extensively to enhance life and reduce size of batteries in devices like cellular phones and laptops, etc.	**Energy audit:** Chapter 5 shows the path for rapid implementation of optimum EE&C technologies and techniques through energy audits and appreciation of their recommendations. It is the essential tool for sustainability of metal industries to enact any ETSs in the EU (European Parliament, 2015), Energy Savings Certificates (ESCerts) trading under PAT scheme in India, T-1000, T-10000 and 10-key projects in China as mentioned in Chapter 2, or energy review in ISO 50001 globally.
5	Manufacturing plants and machinery including boilers, turbines, heat-exchangers, furnaces, pumps, motors, conveyors, drives, etc., where steel, copper and also some other non-ferrous metals are used.	**Implementation of EE&C projects:** Chapter 6 deals with salient features of implementation of EE&C projects, the most vital part to achieve sustainability in metal industries. The chapter also explains how M&V protocol plays a very important role in sustaining the energy savings, as well as linking the financial savings from energy savings projects to the balance sheet in BOT and BOOT models of ESCO and successful performance contracting for sustainability in metal industries.
6	Countless other applications: metals are increasingly being used now in newer applications, for example, packaging (aluminum), containers and components replacing plastics (to reduce plastic pollution) and as trace elements in modern medicine.	**Energy management and monitoring systems:** a vibrant energy management and monitoring system in metal industries helps to check conformity with acts and policy guidelines, and invoke expensive activities such as EE&C initiatives and audits—whether implementation of their recommendations lead to desired benefits—and even automation and remote control with modern IT, machine learning, the IoT, and so on are discussed in Chapter 7. All these are necessary for sustainability of the metal industries.

FIGURE 8.11   Innovation at work at Goldcorp Inc.'s all electric Borden Mine.

Source: UNFCCC (2018).

8.8   CONCLUSION

Sustainability in metal industries will remain a challenge, with a high scope to the stakeholders to succeed through ever-evolving procedures, briefly discussed in this chapter. Table 8.4 clearly indicates that the sustainability in metal industries is synonymous to sustainability of modern civilization and calls for a holistic approach and specific solutions as discussed in eight chapters. A SWAT analysis in brief can be placed here with an intention to bridge the gap between importance, opportunities and techniques of EE&C in Metal Industries and its upcoming technologies, innovations and future direction. Strength is only 5–10% energy needs when produced from recycled metals with compared to the high energy intensity of metal industries which can be said to be the backbone of modern civilization (Chapter-1). Lack of coordination at policy level among different verticals of technologies in practice across the globe are the significant weakpoint (Chapter-2). Successful demonstration of breakthrough and emerging technologies and setting of the targets begun to be set by larger player countries at the world forum for 2050s are the most notable opportunities (Chapters-3, 4, 5, 6 & 7). Major threat is the actual action limping behind and present achievement being much disproportionate to the future commitments. The sequential approach in the eight chapters would help readers to march forward with confidence towards the goal of higher EE&C in metal industries for its sustenance and that of human civilization (Chapter-8).

REFERENCES

Accenture (2020); www.accenture.com/_acnmedia/PDF-121/Accenture-Navigating-Impact-COVID-19-Mining-Metals.pdf: Responding to COVID-19 Navigating the impact on the mining and metals industries, Accenture, 2020.

CDP (2021); www.cdp.net/en/companies-discloser: Environmental transparency and account-ability are vital to tracking progress towards a thriving, sustainable future, CDP, 2021

CFI (2021); https://corporatefinanceinstitute.com/resources/knowledge/other/esg-environmental-social-governance/: The framework for assessing the impact of the sustainability and ethical practices of a company; CFI, 2021.

Eurometal (2021): https://eurometal.net/thyssenkrupp-quantifies-iron-loss-from-blast-furnace-interruption/; Thyssenkrupp quantifies iron loss from blast furnace interruption, Eurometal, 2021.

Eurometaux (2015); https://eurometaux.eu/media/1523/full-lt-sustainability-framework-document-approved-1.pdf: Our Metals Future: The metals industry's 2050 vision for a Sustainable Europe, EUROMETALAUX, 2015.

Eurometaux (2020); https://eurometaux.eu/media/2052/paper-covid19-impacts-on-metals-industry-final.pdf: COVID-19: Impacts on the European Metals Industry, Eurometaux, 2020.

European Parliament (2015); www.europarl.europa.eu/doceo/document/A-8-2015-0309_EN.pdf: Report on developing a sustainable European industry of base metals (2014/2211(INI)), Committee on Industry, Research and Energy, European Parliament, 2015.

European Union (2010); https://cordis.europa.eu/project/id/515960: Ultra-Low CO2 steel-making; European Union, 2010.

European Union (2019); NON-ENERGY MINERAL EXTRACTION IN RELATION TO NATURA 2000 CASE STUDIES, October 2019, European Union Publication. https://ec.europa.eu/environment/nature/natura2000/management/docs/NEEI%20case%20studies%20-%20Final%20booklet.pdf

European Union (2021); https://ec.europa.eu/clima/policies/eu-climate-action/2030_ctp_en: 2030 Climate Target Plan, European Union, 2021.

Global CCS Institute (2017) CCS: A necessary technology for decarbonising the steel sector; Global CCS Institute, 2017; https://www.globalccsinstitute.com/news-media/insights/ccs-a-necessary-technology-for-decarbonising-the-steel-sector/

GRI (2022); www.globalreporting.org/: GRI Standards, 2022.

He, Kun, Li Wang, and Xiaoyan Li (2020). "Review of the Energy Consumption and Production Structure of China's Steel Industry: Current Situation and Future Development" Metals 10, no. 3: 302. https://doi.org/10.3390/met10030302

HSE (2001); https://www.hse.gov.uk/pubns/web34.pdf; The explosion of No. 5 Blast Furnace, Corus UK Ltd, Port Talbot; Health and Safety Executive (HSE), Government of UK, 2001.

ISO (2022); https://www.iso.org/developing-sustainably.html: Developing Sustainability, 2022.

Lèbre, É., Stringer, M., Svobodova, K. et al. The social and environmental complexities of extracting energy transition metals. *Nature Communications*, Volume 11, No. 4823, 2020. https://doi.org/10.1038/s41467-020-18661-9

Mazin Obaidat, Ahmed Al-Ghandoor, Patrick Phelan, Rene Villalobos and Ammar Alkhalidi (2018); Energy and Exergy Analyses of Different Aluminum Reduction Technologies; Sustainability 2018, 10, 1216; doi:10.3390/su10041216; www.mdpi.com/2071-1050/10/4/1216/pdf.

Michel Y. Haller, (2020); https://reader.elsevier.com/reader/sd/pii/S2590174519300157?token=ACE8DFFA63801C8DD2467C87C8B7BF0510AEC77F8E9F53BC0A2226B729ACE591BEF111B5FAEC73933D4EBA4F4A87B3D7: Michel Y. Haller, Seasonal energy storage in aluminium for 100 percent solar heat and electricity supply, Energy Conversion and Management: X 5 (2020) 100017.

Pauliuk et al. (2013); https://doi.org/10.1016/j.resconrec.2012.11.008: Stefan, Pauliuk, Wang, Tao, Müller, Daniel B. Steel all over the world: Estimating in-use stocks of iron for 200 countries. *Resources, Conservation and Recycling*, Volume 71, Pages 22–30, ISSN 0921-3449, 2013.

Roosa (2010); Roosa, Stephen A. *Sustainable Development Handbook*, 2nd Edition, The Fairmont Press, 2010.

Sengupta et al (2018); Sengupta, P., Dutta, S.K., Choudhury, B.K. Waste heat recovery policy. In: Gautam, A., De, S., Dhar, A., Gupta, J., Pandey, A. (eds.), *Sustainable Energy and Transportation. Energy, Environment, and Sustainability.* Springer, 2018. https://doi. org/10.1007/978-981-10-7509-4_11

S&P Global (2019); Prepared by Krishnakumar Vishwanathan as reported at ESG Industry Report Card: Metals and Mining; S&P Global, June 3, 2019. www.spglobal.com/_media/documents/spglobalratings_esgindustryreportcardmetalsandmining_jun_03_2019-003-.pdf

S&P Global (2020); www.spglobal.com/ratings/en/research/articles/200722-environmental-social-and-governance-our-updated-esg-risk-atlas-and-key-sustainability-factors-a-companion-11583314: "Our Updated ESG Risk Atlas and Key Sustainability Factors: A Companion Guide," S&P Global, 2020.

Tata Steel (2020) HISARNA – Building a sustainable steel industry; Tata Steel, 2020; www.tatasteeleurope.com/ts/construction/sustainability/sustainable-construction-in-steel#ccs:Hisarna https://www.tatasteeleurope.com/sites/default/files/TS%20 Factsheet%20Hisarna%20ENG%20jan2020%20Vfinal03%204%20pag%20digital.pdf

Thomas Koch Blank (2020); https://energypost.eu/hybrit-project-sweden-goes-for-zero-carbon-steel/: HYBRIT project Sweden goes for zero-carbon steel.

UNCTAD (2016); https://unctad.org/system/files/official-document/ciiisard78_en.pdf: 'Enhancing the role of reporting in attaining the sustainable development goals: Integration of environmental, social and governance information into company reporting', UNCTAD, 2016.

UNCTAD (2017); https://unctad.org/system/files/official-document/ciiisard81_en.pdf: 'Enhancing comparability of sustainability reporting – Selection of CIs for CR on SDGs', UNCTAD, 2017.

UNCTAD (2018); https://unctad.org/system/files/official-document/ser-rp-2018d1_en.pdf: Reporting on the Sustainable Development Goals – A Survey of Reporting Indicators, UNCTAD, 2018.

UNFCCC (2018); https://unfccc.int/sites/default/files/resource/GCA_Yearbook2018_Annex04_ Industry_Snapshot.pdf: Industry Sector Snapshot Mining and Metals, UNFCCC, 2018

UNFCCC (2021a); https://unfccc.int/climate-action/race-to-zero-campaign#eq-4: How to join Race to Zero, UNFCCC, 2021

UNFCCC (2021b); https://unfccc.int/news/un-welcomes-us-announcement-to-rejoin-paris-agreement: UN Welcomes US Announcement to Rejoin Paris Agreement, UNFCCC, 2021.

UNGLOBALCOMPACT (2021); www.unglobalcompact.org/: A global movement – discover ways to engage, UNGLOBALCOMPACT, 2021.

United Nations (1987); https://sustainabledevelopment.un.org/content/documents/5987our-common-future.pdf: "Report of the World Commission on Environment and Development: Our Common Future", United Nations, 1987.

United Nations (2015); https://sdgs.un.org/2030agenda: "Transforming our world: The 2030 Agenda for Sustainable Development", United Nations, 2015.

World Economic Forum (2021); https://reports.weforum.org/toward-the-circular-economy-accelerating-the-scale-up-across-global-supply-chains/from-linear-to-circular-accelerating-a-proven-concept/#:~:text=A%20circular%20economy%20is%20 an,regenerative%20by%20intention%20and%20design.&text=First%2C%20at%20 its%20core%2C%20a,cycle%20of%20disassembly%20and%20reuse: From linear to circular – Accelerating a proven concept, Pp - ; World Economic Forum, 2021.

World Energy Council (2021); www.worldenergy.org/transition-toolkit/world-energy-trilemma-index: World Energy Trilemma Index, World Energy Council, 2021.

World Steel Association (2015); www.worldsteel.org/en/dam/jcr:00892d89-551e-42d9-ae68-abdbd3b507a1/Steel+in+the+circular+economy+-+A+life+cycle+perspective.pdf: Steel in the Circular Economy – A life cycle perspective, World Steel Association, 2015.

World Steel Association (2020); www.worldsteel.org/en/dam/jcr:6d73d7fa-9739-439e-96d8-b2d57b32951c/Indicator%2520data%2520report%25202020.pdf: Sustainability Indicators, World Steel Association, 2020.

Index

For Product Safety Concerns and Information please contact our EU
representative GPSR@taylorandfrancis.com
Taylor & Francis Verlag GmbH, Kaufingerstraße 24, 80331 München, Germany

www.ingramcontent.com/pod-product-compliance
Lightning Source LLC
Chambersburg PA
CBHW060407220326
41598CB00023B/3046